Encounters
with Euclid

Also by Benjamin Wardhaugh

*Poor Robin's Prophecies: A Curious Almanac,
and the Everyday Mathematics of Georgian Britain*

*A Wealth of Numbers: An Anthology of
500 Years of Popular Mathematics Writing*

*Gunpowder & Geometry: The Life of Charles Hutton,
Pit Boy, Mathematician and Scientific Rebel*

Encounters with Euclid

How an Ancient Greek Geometry
Text Shaped the World

Benjamin Wardhaugh

Princeton University Press
Princeton and Oxford

Published in the United States and Canada in 2021 by Princeton University Press
41 William Street, Princeton, New Jersey 08540

press.princeton.edu

Originally published in the English language in 2020 by HarperCollins Publishers Ltd. under the title *The Book of Wonders: The Many Lives of Euclid's* Elements

Library of Congress Control Number 2021930241
ISBN 978-0-691-21169-5
ISBN (e-book) 978-0-691-21981-3

Typeset in Sabon LT Std by Palimpsest Book Production Limited, Falkirk, Stirlingshire

Printed on acid-free paper. ∞
Printed in the United States of America

1 3 5 7 9 10 8 6 4 2

For my parents

Contents

CONTENTS

IV: SHADOW AND MASK

CONTENTS

Prologue

Alexandria. Alexándreia. The reign of Ptolemy, first of the Alexandrians. Perhaps the tenth year of his reign: a little after 300 BC.

Arrive by sea, the Egyptian sun on the water. Cross the harbour; enter the city. Building upon building. Through the Gate of the Moon and up the boulevard, the Canopic Way. White marble; dust and hammering: building work everywhere. The grandest of cities. You can drive your carriage up the streets if you want to. Paved streets and white-faced buildings. The sea murmuring.

To the crossroads; turn left onto the Street of the Sema: long and cool, catching the wind. Into the palace quarter. Temples; museum; library.

One of the workers in the famous cultural quarter is a man named Euclid. One of his books, the *Elements of Geometry*. When Ptolemy's glorious Alexandria is dust, it will live.

For twenty-three centuries the *Elements of Geometry* has been changing the world. A compendium of facts about space and its properties – lines and shapes, numbers and ratios – it has drawn countless readers into its limitless world of abstract beauties and pure ideas. And it has been on the most incredible journey. Few artefacts survive the collapse of the culture that produced them; few texts survive the abandonment of the language in which they are written. The *Elements* has survived both. Indeed, it seems to have positively thrived on its transplantation into a series of fantastically diverse situations. Its very austerity seems to have enabled readers to find in it qualities that made it interesting to them, characteristics that gave it meaning in their time and place.

The sculptors of the west facade of Chartres Cathedral depicted Euclid, and the scholars of Abbasid Baghdad translated his book. An American artist turned his diagrams into art, and an Athenian philosopher wrote a commentary on them. The *Elements* was implicated in the Scientific Revolution: the fateful decision to read the book of nature as if it was written in the language of mathematics.

In Beijing, between August 1606 and April the following year, the scholar Xu Guangqi and the Italian Jesuit Matteo Ricci worked to translate the *Elements*, one of the books Ricci had brought with him from the far west, into the language of the Mandarin scholar. They struggled with terminology, with the structure of the text, with the very different assumptions each brought to its subject matter. They revised the text three times before they were happy to release it for publication.

From May to November 1817, on the other side of the world, Anne Lister set aside her mornings for – in equal proportions – arithmetic and Euclid. By the autumn she had worked through more of the *Elements* than most university graduates.

A thousand years before, at Gandersheim in Lower Saxony, the canoness Hroswitha wrote into one of her plays Euclid's definition of perfect numbers. It formed part of Wisdom's mockery of the emperor Hadrian, who was seeking to torture her and her daughters.

Again and again, new generations in new places have encountered the *Elements* and done new things with it. It has travelled through worlds that could never have been imagined by the Greeks who first wrote and read the text.

What does it mean for a book to live 2,000 years and more? To survive the wreck of the civilisation that produced it? To find readers again and again, in place after place and time after time? How vast a range of meanings do readers have to find in it? How vast a range of readers does it have to find?

Come on a journey and find out.

I
Author

Alexandria

The geometer and the king

Alexandria, around 300 BC.

A dinner, say: a symposium, in the palace quarter, perhaps in the Museum. Ptolemy himself in attendance: general, hero, king, deity. And the talk touches on geometry: why so hard? Why is there no easier way? The geometer – a dusty man, but pert – answers: Majesty, there is no royal road to geometry.

Ptolemy snubbed is one of the irresistible stories. The man had been a childhood friend of Alexander the Great; one of his body-guard. Maybe his illegitimate half-brother. He was a trusted general (the name is said to mean 'warlike'): level-headed but capable of the grand gesture; a man of no nonsense.

He was one of the great survivors, too. In the twenty years of chaos that followed Alexander's death, when many abler men died, Ptolemy played and won. Of all the successors who eventually carved up Alexander's brief, continent-spanning empire, he founded the longest dynasty, the stablest country. He chose Egypt, and he never risked it for a larger realm. Fourteen Ptolemaic rulers followed him, until Cleopatra lost it all at the Battle of Actium 250 years later. The first king, then, of the last Egyptian dynasty. *Basileus* to the Greeks, pharaoh to the Egyptians. Inheritor of 3,000 years of Egyptian kingship and, yes, a god too. In 306 BC he defeated an attack on Rhodes so soundly that altars were erected to him and he acquired the title 'saviour'. By 278 BC there were Ptolemaic games in his honour: four-yearly, like the Olympics.

Ptolemy I Soter.

The geometer, by contrast – a man named Euclid; Eukleídēs – is utterly obscure, historically speaking. The story about the 'royal road' is moonshine, sadly; it is also told about another geometer (Menaechmus) and another king (Alexander), and there is little reason to suppose it really happened. Even his dates – sometime around 300 BC – are the mere surmise of authors writing centuries after him. Unlike the exceptionally well-documented Ptolemy, Euclid left no biographical traces whatever. He founded no dynasty, built no palaces. His legacy was solely intellectual.

But what a legacy. The line of his students at Alexandria outlasted his life. His book outlasted his civilisation.

What was this city, that produced such a man and such a book? Alexandria was the right setting for the *Elements of Geometry*. It was Ptolemy's greatest achievement. Alexander himself decreed the city, on the site of what had been a village and, like a dozen others, it bore his name. He never saw a stone erected, but Ptolemy took it for his capital, moving the Egyptian royal seat from Memphis. It was a Greek *polis* in a profoundly un-Greek world, a new foundation in a land where cities were 2,000 years old. Ptolemy did everything to make it splendid; it had an assembly, a council, its own coinage and its own laws. There were broad boulevards, colonnades, avenues of trees, street lamps. In 322 BC he kidnapped Alexander's corpse and put it on display in his new royal city.

It was indeed the most splendid site for a city, at the meeting place of two continents, just to the west of the mouths of the Nile. It would be a major commercial port for centuries, a strategic military site until the Second World War. Ptolemy laid the foundations of the famous lighthouse: fortress and beacon, it would be one of the seven wonders of the world when it was done, 400 feet high and topped by a statue of Zeus. It stood for 1,500 years. Blessed with such a city, people immigrated from all over the Greek world, and Alexandria became not just notoriously big and splendid but notoriously crowded and cosmopolitan, with Greeks, Macedonians, Egyptians, Jews, Syrians and more, its streets teeming like ant heaps. Within a few generations it would have more than a million inhabitants.

As well as town planning, and the sheer erection of building upon building, Ptolemy involved himself in cultural policy: and with his characteristic effectiveness. To make himself a convincing Egyptian pharaoh, he indulged in sculpture to match, and devised a new cult of 'Serapis', a blatantly invented figure with a hybrid iconography. Like all his achievements, it endured: the temple, the Serapeum at Alexandria, stood for 600 years.

And to please the Greek mind and heart there were pageants, festivals, and a palace with tapestries the gods would prize. As one contemporary put it, Alexandria had 'wealth, wrestling schools, power, tranquillity, fame, spectacles, philosophers, gold, youths, the sanctuary of the sibling gods . . . the Museum, wine, every good thing [one] could desire'.

All of which was invaluable in projecting Greek power and an idea of Greekness in a profoundly alien environment; in saying, this is what we do in the great Greek world. This is our right to rule.

And so, the Museum. The Mouseion: the shrine of the Muses. It was royally subsidised, with scholars in discipline after discipline. Their head was a priest of the Muses, and they included poets, grammarians, historians, philosophers, doctors, natural philosophers, geographers, machine-builders, astronomers, and of course geometers. It was partly Ptolemy's own impulse, partly the creation of Demetrios of Phaleron, a famous pupil of Aristotle brought over from Athens to oversee the creation of the new institution. It had courts, covered walks and gardens, a dining hall and an observatory. The staff numbered perhaps forty, and they spent their time researching, writing and sometimes teaching. They held learned symposia, some attended by the king. They were a remarkable collection of people, sometimes compared a little sourly with the collection of animals that Ptolemy also founded: 'well-fed bookworms, arguing endlessly in the Muses' birdcage'. If there were bookworms, there were certainly books, too; the library at Alexandria would become the most famous in the world, though it seems to have been organised a little later, under Ptolemy's son.

All of which is how the famous Greek mathematician ended up working in Egypt. Was Euclid another part of Ptolemy's collection, someone brought in to swell the ranks of the Museum? It is not certain whether he was native to Alexandria or an immigrant, although at this early date in the city's life the latter is much the more likely. An immigrant from where? His austere prose gives no hint of an accent: unlike Archimedes, in the next generation, who wrote in the Doric dialect of Syracuse.

What came to Alexandria in Euclid's person (and perhaps in the persons of other mathematicians; it is not clear whether or not he was the only one) was a well-established tradition of Greek geometry. Greeks liked to have things to think about. They liked hobbies. Some Greeks raced chariots, some talked about philosophy, some occupied themselves with politics. From perhaps the late fifth century BC onwards, some did geometry.

What was it like? Perhaps it is clearest to think of Greek geometry as an outgrowth of the Greek love of talk, of disputation. For its geometry was nothing if it was not a performance.

Draw a line, a square, a circle. Reason out loud as you draw; play to the inevitable audience. From such beginnings the long-lived game of geometrical reasoning took shape. The figure of the geometer drawing in the sand remains part of the image of ancient Greek mathematics to this day: raking in the 'learned dust', as the Roman orator Cicero put it. He memorably evoked Archimedes as a man of 'dust and drawing-stick'. (Though have you ever tried to draw a detailed diagram in dry sand? Pieces of clay, wax tablets or, for showing to a larger audience, wooden boards seem more probable.)

The number of Greek mathematicians was never very large, and they had to write their ideas down to preserve what they had found out about their lines and circles; there weren't enough of them, it seems, for a purely oral transmission to stand a chance. So a genre was born: a special style of mathematical writing. That style would come to define mathematics in the West for more than two millennia, as constraining as any poetic metre, and as long-lived. Its components were the statement (of something to be proved); the diagram, with its points labelled with letters; and a chain of reasoning from things already known to things newly proved. That chain ended with the proposed, intended result, and the section – the 'proposition' – was rounded off with the note that 'this is what was to be proved': *hóper édei deîxai*; *quod erat demonstrandum*; QED. In some cases it was 'what was to be drawn'. Here is an example:

How to draw an equilateral triangle

Start with any straight line; call its ends A and B.

Now draw two circles, each with its radius equal to that line: one centred at A, the other centred at B.

The two circles will cross at two points. Pick one of them and call it C. Now join up A, B and C. They make an equilateral triangle.

Why?

Because of the way the distance from A to B was used to find C. C is that same distance from A, and C is also that same distance from B. In other words, all three of the triangle's sides – AB, BC and CA – are the same length. So it's an equilateral triangle. Which is what was to be drawn.

The same ancient sources that report when Euclid lived also report that there were written collections of geometrical learning in Greek by perhaps 400 BC: a century before him. They are quite informative about the subjects involved, and even some of the specific results and techniques. The written matter itself has not survived, though, throwing it all into doubt. There is a strong temptation to make up genealogies for mathematical ideas when real evidence is lacking. So yes, maybe the study of the circle was

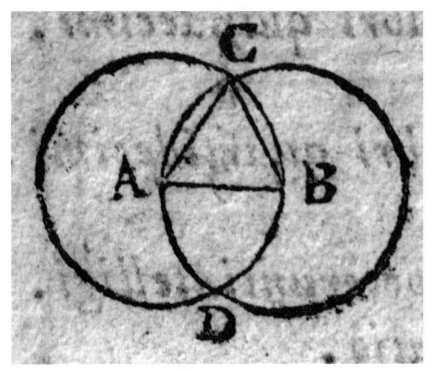

Constructing an equilateral triangle.

done by the Pythagoreans; likewise the work on numbers and their properties. Very probably the work on ratios was done by a geometer called Eudoxus early in the fourth century BC. Some work on the regular solids originated with another, named Theaetetus. The claim that there were, before Euclid, well-developed books called *Elements of Geometry* seems distinctly more doubtful.

What was Euclid's role, then? He took all the easier material known to the Greek geometers of his day and put it all together in a single book. He organised it, both on the large scale and the small. Certainly he added some new matter of his own, though no one can now confidently identify what was new and what was not. Historians continue to argue – and they always will – about how much was compilation and how much composition in Euclid's book. His was a work of collection, like Ptolemy's; Euclid the

Museum artefact became himself a curator, the *Elements* a smaller museum.

Yet this museum contained a world. It displayed the geometrical style of prose, in a relentless, ritualistic tramp of proposition after proposition: 400 of them, arranged in thirteen 'books' or chapters. Every verb was perfect, imperative and passive: 'let a circle have been drawn'. There was something hypnotic about it; something infinitely calm. The book began with definitions: what was meant by a line? A point? A circle? It continued with the simplest manipulations of lines and shapes in two dimensions: how to draw different kinds of triangle; how to divide a line or an angle in two. The fact that, in a triangle, any two sides add up to more than the third. The Epicurean philosophers thought that last piece of geometrical information was 'evident even to an ass' since 'if straw is placed at one extremity of the sides, an ass in quest of provender will make his way along the one side and not by way of the two others'.

Euclid did not care how obvious any of it was. He arranged and exemplified a toolbox of the basic techniques and results that he had inherited: ways of arguing, ways of proving; facts that geometers commonly assumed or used but seldom proved in full. At the end of the first book he placed 'Pythagoras' theorem'. Draw a triangle, with one of its angles a right angle. Using its shortest side as a base, draw a square, whose side is the same length as that side of the triangle. Repeat the procedure with the triangle's two longer sides, so that you end up with three squares of different sizes, resting flush against the three sides of the triangle. Now, it turns out, the areas of the two smaller squares will add up to that of the larger one: a startling fact, not evident to any ass, which Euclid proved in his characteristically meticulous manner.

And so the ideas and the diagrams became harder and more complicated throughout the book. There were purely geometrical sections: a description of how to draw a regular pentagon or hexagon inside a given circle, for example. Parts of the book dealt not with geometry but with numbers and ratios, ranging from the most basic facts ('if an odd number is multiplied by an odd number, then the product is odd') to a procedure for finding the mysterious 'perfect' numbers, equal to the sum of their divisors.

Finally, Euclid turned to three-dimensional shapes. The last three books of the *Elements* – books 11, 12 and 13 – were concerned with spheres, cones and cylinders, with cubes and cuboids and with the regular polyhedra. These were beautiful solids whose faces were all the same regular polygon: triangles, squares or pentagons. There were just five regular polyhedra: the tetrahedron (four triangular faces), the cube, the octahedron (eight triangles), the dodecahedron (twelve pentagons) and the icosahedron (twenty triangles). He showed how to make such shapes, starting from, say, a given triangle or a given circle; he showed how to find their surface areas and their volumes. Euclid's explorations in these final books were frequently ingenious, and sometimes applied an almost unbelievable amount of lateral thinking. Despite its gentle beginnings, and its incorporation of a lot that anyone could understand, the *Elements* as a whole was a virtuoso performance; a road that only the keenest of geometrical minds could follow all the way.

It added up to more than 20,000 lines of text in Greek. Euclid was careful, but he was no superman, and both joins and slips were sometimes visible. A few definitions (oblong, rhombus, rhomboid) seemed to have been carried across from older sources but never actually used. Quite a number of terms were used that were, conversely, not defined earlier on. Some words were ambiguous. A surprising amount was taken for granted about the properties of points and lines that Euclid never explicitly set out in his assumptions. Some propositions were merely special cases of others; some propositions were, strictly speaking, unnecessary because they were the mere logical consequences of others. But despite such wrinkles, the *Elements* was a fine, even an awe-inspiring monument to all that had been done in Greek geometry so far.

Euclid was by no means the author of just one book. The order of events is not clear, but it is certain that he wrote more. There were perhaps four other books on special topics in elementary geometry, and there were books on the applications of mathematics – to music, astronomy, optics and more – as well. In total nearly a dozen books are mentioned in the early evidence; eight actual texts survive, though most are disputed by historians.

Back in seething Alexandria, where the building work is still going on and the streets are ever more crowded. By the end of Euclid's life the great lighthouse at Pharos has been built (did the architects consult Euclid? It would be intriguing to think so); the library and Museum are nearing completion and the palace complex is grander than ever. The *Elements* is finished: thirteen rolls of papyrus covered with neat columns of text and diagrams. And Euclid is still teaching, still taking on new students.

One beginner is impatient, like the king before him. After he has understood the first proposition he bursts out, 'What is my profit now that I have learned that?'

A glance of contempt, or perhaps of pity. Euclid calls for a servant. 'Give him threepence, since he must always make a profit out of what he learns.'

Another romantic legend, perhaps: one that circulated later in Greece, during its period of Roman domination. Like the 'royal road' story, it helped to protect Euclid from the whiff of servility, of sycophancy that hung over anyone connected with Ptolemaic Alexandria and its institutions. It preserved and dramatised the idea that geometry was a leisured, cultured pursuit, part of the life of the mind. Not a profitable trade, but pure, truthful and beautiful for its own perfect sake.

Some 350 geometrical propositions in the driest of styles. It's an

odd thing, to have become one of the most enduring cultural artefacts of the Greek world. Ptolemaic Alexandria is largely dust
today; a few wrecked statues are dug from the ground or pulled
from the sea from time to time, but the splendour is all gone.
Ptolemy's dynasty died with Cleopatra. The library is scattered.
But the books – the *Elements* among them – the books lived.

Elephantine

Pot shards

Elephantine island, Upper Egypt, in the reign of Ptolemy III (r.246–221 BC; grandson of Ptolemy I). A Greek garrison at the end of the world. One man, off duty, is writing.

He grabs the nearest thing to hand to write on: a few broken pieces of pot. Scratch, scribble. A quick diagram, a few lines of text. His hands are confident; only slightly less confident is his memory of the mathematics. Is this how it worked? Or this? Ah, that's right.

And, his mind and hand refreshed, the potsherds, the cheapest available writing material – perhaps the cheapest possible – go into the rubbish heap where they belong.

Euclid's original manuscripts do not survive, nor anything like them. The papyrus on which he wrote is durable enough, in the right conditions. Scrolls hundreds of years old were not terribly unusual in the ancient world, and they could remain smooth, pliable and legible for much longer. A story is told of a museum curator who used to display the strength and flexibility of papyrus by blithely rolling and unrolling an Egyptian sheet 3,000 years old (this was in the 1930s, when attitudes to museum artefacts were perhaps less reverent than today).

In the right conditions, that is. Most conditions are not right. If it gets too wet, papyrus rots; too dry and it crumbles. Insect larvae like papyrus, and the worms destroyed many a literary reputation in the ancient world. So did the rats. Plus, the long rolls tore easily and were thrown away when they did. The upshot is that large or complete papyri surviving from the ancient world are extremely rare. What more often survive are fragments: discarded rolls, pieces reused to make mummy cases, pieces recovered from rubbish dumps or ruined houses. Rough, dark and brittle with age, nearly all are from provincial locations in Middle and Upper Egypt, where the dry conditions preserved them. Finds have come from cemeteries along the Nile Valley and in the Faiyum Oasis, and from certain villages. From the big towns, by contrast, there is next to nothing: Alexandria itself, having a high water table, has no preserved papyri at all.

For all that, there are a lot of papyrus fragments. People have been systematically digging them out of the ground since the mid-nineteenth century, and hundreds of thousands are now amassed. And, yes, some of them contain fragments of Euclid's *Elements*. Seven, in fact, totalling about sixty complete lines of the text and another sixty fragmentary lines.

What parts of the *Elements* do they preserve? They include, written around 100 BC: three propositions from book 1, with one summary proof (these come as citations in a philosophical treatise preserved – carbonised – in Herculaneum by the eruption of Vesuvius in AD 79: an exception to the normal generalisations about papyrus survival). An enunciation from book 2, with a rough figure, written in the Egyptian city of Oxyrhynchus around AD 100. Parts of two more propositions from book 1, written at Arsinoë (modern Faiyum) in the second half of the second century AD. A second-century copy of three figures and enunciations from book 1, carefully written with ruled diagrams. And a schoolteacher's or pupil's copy of the ten opening definitions, made in the third century AD.

It is not much: these are small pieces from the easy parts of the book, in one case from its very beginning. But they do reveal something about the way the *Elements* was spreading. It did not just stay in Alexandria: already, by the first few centuries after its composition, it – or parts of it – was being copied out by people hundreds of miles away around the Greek-speaking world. It was moving out from the cultural centre to the provinces.

Euclid's *Elements* will have been published in the ancient sense: sent to a scribal copying house which produced multiple copies for sale. But most of the papyrus fragments are not from those copies; only the Faiyum fragment looks like the work of a professional scribe. Instead, they bear witness to the activity of individuals copying out parts of the text for their own use, teaching or learning.

So, the writers of these papyrus fragments represent the 'public' for Greek geometry: a tiny minority, in a world in which the literate themselves were already a minority. These were people who understood geometry, who accepted and shared its conventions, who knew enough of the basics and the methods to comprehend Euclid's book. Their needs surely shaped what was written and how it was written. The very packaging of mathematics in a self-contained written form already presumes that they existed. But nothing more is known about them.

And this evidence can tell only about those places where it was dry enough to preserve papyrus fragments: for the rest of the Greek world – the islands and the mainland north of the Mediterranean, for instance – the lack of evidence reveals nothing, positive or negative. Surely the *Elements* went to Athens, for example: but it is centuries before there is evidence for that.

As well as papyrus there was a cheaper writing surface still: ostraka or pot shards. Literally, broken pieces of pot: waste, and therefore free. Ostraka were used in Egypt before the Ptolemies and up to the end of antiquity, in Athens from the seventh century BC: they were written on in ink or simply scratched, to form pictures or writing in Hieratic, Demotic, Greek, Coptic or Arabic as the case might be. Schoolboys, soldiers, priests and tax collectors all used them. (They were also used as voting tokens: if the word sounds familiar it is because 'ostracism' was a procedure for expelling a man from the country for ten years on suspicion of disloyalty: the votes were written on ostraka. It happened at Athens for most of the fifth century BC, and in other Greek cities too.)

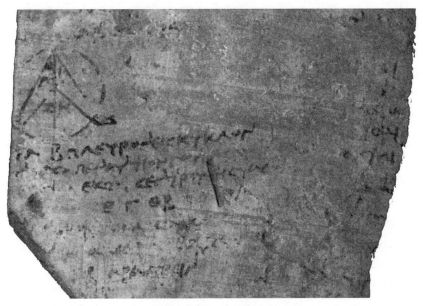

A Euclidean ostrakon.

A set of ostraka bearing geometrical writing, written at Elephantine, were preserved by the random sieve of history. They were dug up by German archaeologist Otto Rubensohn in the winters of 1906 and 1907, and they are now held with the papyrus collection at Berlin; their contents were transcribed and published in the 1930s. They are the oldest surviving physical evidence for any part of Euclid's *Elements*.

Elephantine is more than 500 miles south of Alexandria: an island in the Nile at the northern end of the first cataract. In the third quarter of the third century BC, when the ostraka were written, it was the frontier of the Ptolemaic kingdom. Traditionally the 'ivory island' or 'elephant island', whose settlement stretched back into prehistory, it was the capital of the first Upper Egyptian administrative district or 'nome', controlling trade with the quarries of the cataract region and the trade route to Nubia. It was a garrison town, far removed from the centres of Greek culture, under threat from brigands.

It had temples, priests, fairly elaborate housing and a degree of bustle; by the Byzantine period the town had a public camel yard. But the documents that have survived – papyri again, mainly – are dominated by the characteristic anxieties of soldiers settled far from their widely diverse homelands. In the third century BC there were men at Elephantine from Greek cities and islands as far afield as Crete and Rhodes as well as Alexandria and the mainland at Euboea and Phocis: a veritable Homeric catalogue of soldiers. They kept aloof from the native Egyptians, referring to their town as 'the fortress', and the papyri show them making wills, marrying, appointing guardians, providing accounts to their superiors or petitioning them for justice. It seems a most unlikely setting for the earliest Euclidean evidence.

The ostraka – six of them, one clearly broken on all sides – bear a text concerned with constructing a regular polyhedron. It relates to propositions 10 and 16 in book 13 of the *Elements*: that is, from very near the end of the book. In those propositions a pentagon, hexagon and decagon are employed to build up an icosahedron: a regular solid with twenty sides, all of them equilateral triangles. The inevitable diagram is clearly present on one of the ostraka, with its letter labels, and the shards give a clear sense of someone doing what every Greek geometer did: draw a picture and tell a story about it.

The propositions in the *Elements* depend on one another in an elaborate, tree-like structure, each one referring implicitly to many that have come before. To be confident in something from this late in the book, a person would need to have studied much of what came before it: but the ostraka provide no direct evidence for that. The text is in a confident, flowing handwriting, too: that of an experienced writer whose grammar and spelling were unhesitating and correct.

Who was the writer? It is most frustrating not to know. Serafina Cuomo, historian of ancient mathematics, remarks of these ostraka that 'while their contents denote a high level of education, both the humble material and the location (a remote outpost in the heart of "Egyptian" Egypt) seem to jar with that conclusion'. Priest, teacher, soldier or camp follower? It will never be known whose were the hands that scratched out this earliest surviving piece of Euclideana.

There is a further twist. The matter discussed on the ostraka came straight from *Elements* book 13, but their text is not the text that has been transmitted as part of the book. It is the same diagram, and the same ideas, but the words are not exactly the same. That is also true, albeit to a lesser degree, of the other early fragments that survive from the *Elements*, on papyrus: their versions of the text do not exactly match what has been preserved in later, more complete versions of the book.

This reinforces the sense that Greek geometry was essentially a performance, consisting of drawing a diagram and talking about it, to oneself or to an audience. What was written down was a transcript of the performance, an aide-memoire, a skeleton or a set of prompts, together with the finished, static version of a diagram that, in use, had been dynamic and growing. Those written traces could be of use for private study or a help for a teacher, who would need to perform the same proof several times. They could also, as in a book like the *Elements*, transmit the idea of the proof or construction to people far away in time or place.

And, as a result, reading a geometric proof is not like reading a novel or a poem. You can only really follow it by recreating the original live performance; by picking up a pen and some scrap paper, papyrus, or a shard of pot, and constructing the diagram as you read, so as to see how it grows.

So the written forms of Greek geometric propositions were not so much something one would learn and copy slavishly as prompts that said: here is something interesting; try it yourself. The *Elements* was not a dead repository of facts but a support for learning and practice; an invitation to perform for oneself, in the same way that rhetoric textbooks aimed to prepare students for rhetorical performance. With that in mind it becomes less surprising that the early fragments of the *Elements* surviving on pot and papyrus have quite 'wild' versions of the text. This set of ostraka, in particular, should probably be read in this way: as an attempt to recreate something the writer had read or seen performed.

It is fitting that the earliest evidence for the Euclidean *Elements* is so enigmatic, so shakily related to the *Elements* themselves, and so, literally, fragmented. The text and its ideas would travel about as widely as it is possible for a cultural artefact to travel, but they would be much changed by the journey; and, what is more, it is not clear that they were ever simple, single and stable, even at the very beginning. Euclid was not a master but a muse, an inspiration:

he did not just reveal facts but offered a set of tasks. His readers knew they could always go deeper and create more, because although the *Elements* had already done everything, everything was still to be done.

Hypsicles

The fourteenth book

Back in Alexandria, and still under Ptolemy III. Another legend, this one reported by Galen, the Roman physician.

Ptolemy III's interest in collecting ancient books mounts to an obsession, a mania, and he has ships arriving at the harbour searched for texts that might be of interest to the Alexandrian library. On one occasion he receives, on loan from Athens, the precious scrolls containing the plays of Aeschylus, Sophocles and Euripides, depositing the fantastic sum of fifteen silver talents as security on the understanding that he will have copies made and return the originals.

He has copies made, indeed, and on the best papyrus. But it is those copies that he sends back to Athens, retaining the originals for his library. The Athenians can do nothing; they had accepted the silver on the condition that it would be theirs if the scrolls were not returned. They keep the copies, and they keep the money.

Ptolemy I (who learned there was no royal road to geometry) had abdicated in 284 BC in favour of his son, but the cultural institutions he started continued to grow, as did the great, the fabulous city of Alexandria. Its library became the largest collection of books in the world, an unequalled symbol of literary culture, a vast stocktaking of Greekness that surpassed anything to be found in Greece itself. The Ptolemies' hunger for books became a legend; so did the lengths to which they would go to secure good copies of important works: as in the fabulous story quoted above, where Ptolemy III in the next generation paid a king's ransom for the original Athenian texts of the great tragedies. As well as plays, histories and epics, though, the library had books on cookery, magic and fishing: no subject was excluded.

Meanwhile in mathematics, Euclid's *Elements* remained the acknowledged toolbox of standard techniques and results, in which later Greek geometers used to discover all kinds of new things. Where Euclid had dealt with points, straight lines and circles, they dealt with the shapes made by the intersections of cones and planes, and with the solids formed when those intersections were themselves rotated about an axis: weird forms, surely close to the limit of what the human mind can conceive without the aid of algebra or digital visualisation. Yet they used Euclid's terminology and relied on his propositions constantly. As well, they extended the Greek geometrical style itself in new ways that reflected the cultural preoccupations of Alexandria, with twists and turns, elements of surprise and sudden flashes of intellectual light.

Perhaps a generation or less after Euclid, Archimedes (*c*.287–212 BC) earned a reputation as the most brilliant of the Greek geometers, the most skilled at suspense, surprise, and showy unexpected results. One of his best-known surprises was that if you take a sphere and a cylinder with the same height and the same diameter, their volumes are related in the ratio of 2:3.

The great geometer of the next generation was Apollonius (*c*.262–*c*.190 BC); he shared his name, confusingly, with an Alexandrian poet who was head of the library. One ancient report says he studied in Alexandria with the pupils of Euclid (the passage is in fact the same one – the only one – that places Euclid himself in Alexandria). He systematised the study of conic sections – curves derived from slicing up cones – and adopted, too, the Archimedean manner in which beautiful results emerge as surprises at the end of long arguments.

The tradition continued. In 235 BC Eratosthenes, an astronomer and geometer, became head of the library (in succession to Apollonius the poet, in fact): an important moment for the integration of geometry into the culture of the Museum. Eratosthenes would become famous for deducing, correctly, the size of the earth from astronomical observations.

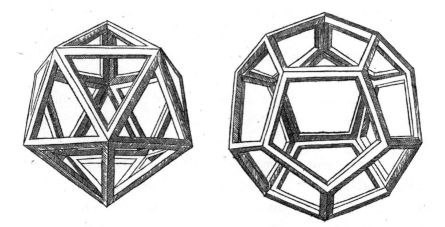

The icosahedron and dodecahedron, in a Renaissance woodcut.

One of the geometrical tricks and puzzles these geometers worked on was this. Take a dodecahedron and an icosahedron of the same size: that is, so that the two spheres into which the two solids would just fit are the same size. The dodecahedron's faces are pentagons; the icosahedron's are triangles. And it turns out – though you really wouldn't expect it – that the pentagons and the triangles are the same size as each other: that is, the circle into which one of the pentagons would just fit is the same size as the circle into which one of the triangles would just fit. It's a fun fact: not one that matters, but a good display of a geometer's mental dexterity. It was admired enough to pass from hand to hand over several generations, and eventually to be annexed to the *Elements*.

It was a geometer named Aristaeus who first made the discovery, and published it in a book on the five regular solids which surely took its inspiration from the discussion that closed the *Elements*. Apollonius treated the problem in a work dealing solely with the dodecahedron and the icosahedron. Next, one Basilides of Tyre came to Alexandria, discussed the matter with a colleague and found Apollonius' account wanting. Apollonius issued a new version of his text, in which he achieved the remarkable feat of showing that the volumes of the equal-sized dodecahedron and icosahedron form the same ratio as do their surface areas. Finally a man named Hypsicles took up the subject and put together all of what he found valid in the previous accounts.

This was in the middle part of the second century BC, perhaps 150 years after Euclid. Much is uncertain, but it seems Hypsicles was an astronomer as well as a geometer; in addition to a now lost book on the harmony of the spheres, and possibly one on numbers, he wrote a work on the rising and setting of the stars. It was in the style of Euclid and showed how to find (approximately) the rising time of a given point in the heavens – say, of a particular star – at a given place on the earth's surface. It was also notable for being the earliest Greek text to divide the circle into 360 parts, a practice that derived ultimately from Babylonia.

Hypsicles responded to Apollonius' revised book with a text of his own on the comparison of the same-sized dodecahedron and icosahedron. He provided new proofs for the main results: that their faces fit into the same circle, and that their volumes have the same ratio as their areas (make a cube of the same size, by the way, and its side and the side of the icosahedron will make that same ratio again). It was a short book, with just five propositions plus three subsidiary results – 'lemmas' – and one alternative proof.

Little is certain about how such things circulated, but it is clear that the treatise of Hypsicles, despite or perhaps because of its recondite subject matter, acquired some currency and was read, cited and admired. But it would suffer a curious fate. A later, anonymous hand attached it to Euclid's *Elements*; where Euclid had written thirteen books, the treatise of Hypsicles was labelled 'book 14' and came to be transmitted as part of the *Elements*. There was some editorial modification at the same time, changing how the 'book' started and ended and also adding an extra proposition to Euclid's own book 13 to strengthen the link. Almost incredibly, the 'fourteenth book' would remain a widely accepted part of the *Elements* for a millennium and a half.

The fantastic collection of books and people at Alexandria, and the mathematical tradition that was becoming associated with the city, interacted in some strange ways and produced some unforeseeable results. The rooms full of boxes full of scrolls made for some unexpected juxtapositions, odd misattributions and unlikely amalgamations. The *Elements* was part of that process, becoming a site of research, controversy and improvement even as it was transmitted from generation to generation.

Theon of Alexandria

Editing the *Elements*

Alexandria in AD 370, or thereabouts. The city remains: Greek, proud and beautiful. But the world is changing: temples are closing, churches are being built. The library is gone, and the Museum's final generation of scholars is at work.

Theon is one of the scholars; his daughter works with him. They teach in the city: geometry, astronomy and philosophy. To teach, you need books, and they prepare new versions of the classic texts: Claudius Ptolemy on astronomy; the Orphic hymns; works on divination; and on geometry, Euclid.

These are long books. The scratch of reed on papyrus is day-long; fingers ache.

Theon comes to *Elements* book 6, proposition 33, and finds that the proposition is not quite complete. He adds a section, extends the result. And so Euclid's *Elements* becomes a little longer and – he hopes – a little better.

Alexandria at its height had been the largest, richest, grandest city of the Greek-speaking world, the Ptolemies the patrons of every art and science. But like all dynasties they had their troubles, and after the Battle of Actium, from 31 BC Egypt was ruled from Rome, made into a province on the Roman model. The city was still rich and splendid; architecturally, it remained perhaps the finest city of the Mediterranean world.

The library lived through it all. The Ptolemies' bibliomania gave it a size that contemporaries could only guess at. One suggested there were nearly half a million scrolls there: surely an impossible number; but whatever the arithmetic in miles of text or acres of papyrus, it was a storehouse of all that was Greek and all that could be amassed and translated into Greek. A most celebrated symbol of its cosmopolitanism was the translation of the Jewish scriptures into Greek during the third and second centuries BC.

But books are fragile, and libraries are fragile. Cultural institutions are not everyone's priority when political instability sets in. Julius Caesar accidentally set the city on fire in 47 BC, and an enduring – and probably true – report says that some books were destroyed, perhaps in seafront warehouses. On the other hand, there was certainly still a functioning library a century later. And the Museum itself continued, even throve at first, under the patronage of the Roman emperors.

But the Roman period saw civil war and riot with at least the usual frequency for a city of such a size: the Alexandrians in fact acquired a reputation for mob violence during their first Roman centuries. In the late third century, civil war reduced the whole of the old palace quarter to 'desert', in the words of one witness, and there was most likely little left of library or Museum after that. It is impossible to be certain. In the evidence that survives, the last employee of the Museum was Theon, who taught mathematics from around 360 until perhaps the end of the century.

Whatever the details of its long demise, the tradition of scholarship at the library had been a crucially important one for the shaping and the sheer survival of Greek literature. Practically every Greek text of any length that survived at all passed through the Alexandrian library.

The library developed a specialism in textual criticism of older literature, studying questions like: what did the authors really write? Which books are authentic? Which lines have been garbled in transmission? It was here that there arose the long-lived metaphor of the corrupt text, the text that needed to be cured, purified of the dirt and disease it had acquired through contact with human beings.

The scholars at the library devised signs for recording which lines of a text were doubtful, and how and where they had emended it. They began to invent marks for section divisions; even line divisions. No punctuation proper; texts were still transmitted in 'continuous writing':

ALLINCAPITALSWITHNOWORDBREAKSACCENTSORPUNCTUA
TIONOFANYKINDANDTHEREFOREREALLYHARDTOREAD.

These techniques and innovations were developed for the study and transmission of literary classics: Homer first, following a tradition of interpretation that went back to the Homeric reciters of classical Athens, and then the Athenian playwrights and others. The scholars at the library had less to say about prose, and less again about scientific texts; there is no evidence that Aristotle's books, for instance, were worked on in this way at Alexandria. To work on mathematics required perhaps a slightly different approach, although the idea of textual purification was the same.

The text of Euclid's *Elements* was already unstable in some ways during its first few centuries; the surviving evidence hints at more. It is quite possible that Euclid himself prepared more than one edition of the *Elements*, which would help to account for some of the variation the book shows in the evidence that survives. It is at least equally possible that zealous editors got to work on the text during the second and first centuries BC, producing versions that differed from the original and perhaps from one another.

What is certain is that by at least the first century AD, and continuing until at least the fifth century, there was a tradition in Greek – and often at Alexandria – of writing commentaries on the *Elements*. Was a definition missing? Was a term defined but not subsequently used? Was a proposition a 'dead end', never used to prove anything else? Did it come in the wrong place, *after* something it was supposed to help prove? All of those things could be pointed out in a commentary, to help teachers and students use the text, to help them avoid stumbling over such minor wrinkles. Commentary of this kind could hardly avoid suggesting changes to the Euclidean text itself, and so it provided raw material to the different but overlapping activity of editing the text: preparing a new version of it for the use of, say, one's own students.

And so to Theon, and the last days of the Museum. He taught, and he dedicated books to his students. He wrote commentaries – most likely based on his lectures – on the second-century astronomical works of Claudius Ptolemy (no relation, as far as is known, of the royal house of the Ptolemies. On his so-called *Handy Tables* he even wrote two different commentaries: one for competent astronomers, and one for those who could not yet understand the reasoning and the mathematics the book contained. He also seems to have written on the interpretation of omens and on the Orphic hymns. A handful of his poems survive, in which he expressed his devotion to the perfect world of the heavens, the gods and the stars.

And he also prepared editions: consistent, improved versions of classic texts for use in his teaching. He certainly worked in this mode on the *Elements* and other works of Euclid, and on the works of Ptolemy such as the *Almagest*, doing what various other editors did to mathematical texts during this period. He cleaned up and smoothed out difficulties. He chose between variant versions of the texts, or combined the variants together to give a text with, sometimes, two alternative proofs for a single proposition. He filled in gaps, real or imagined. While the Homeric editors saw themselves standing outside the epic tradition, mathematicians of later Alexandria tended to behave as though Euclid was a colleague, and they themselves members of a still-living tradition. Instead of textual purity and fidelity ('what did Euclid really write?') they prized correctness, completeness and usability.

Most unusually, Theon actually left a record of one of his specific changes to the *Elements*. In another work – his commentary on Ptolemy – he mentioned it, a certain minor point that 'we have proved in our edition of the *Elements* at the end of the sixth book'. Thus he recorded for posterity that a certain fragment of the text was his, and that that was the kind of thing he did.

The proposition in question says that if you take a slice out of a circle – like a slice of cake, with two straight lines meeting at the centre – then the length of its curved side is in proportion to the angle at the centre. Theon's addition says that the area of the slice is also in proportion to the angle at the centre. It is easy to believe that it is true, and not at all hard to prove.

Theon's throwaway comment has received an enormous amount of attention from scholars of the *Elements* over the years, since it seems to raise the tantalising possibility of seeing beneath the editorial accretions of the ancient world, to the real Euclid beneath. If the type of thing Theon did can be securely described, perhaps later scholars can recognise his modifications elsewhere and reverse them.

Unfortunately, attributing specific modifications to Theon other than this one proves to be a very doubtful business, although many have tried. One person's logical gap is another person's elegant brevity. One person's Theonine interpolation is another person's authentic Euclidean blunder. The evidence – most of it far, far later than Theon, and the subject of endless cross-contamination between Theon's version and the version or versions that preceded him – does not allow anything like certainty.

History, meanwhile, has not been kind to Theon himself. Later ages would have different attitudes to text and authenticity, and would not be too impressed with the kind of modification in which he and his like engaged. Recent historians have characterised his editions as 'trivial reworkings', 'mathematically banal', and his scholarship as 'completely unoriginal'. His biographer in the *Dictionary of Scientific Biography* is frankly devastating: 'for a man of such mediocrity Theon was uncommonly influential'.

Perhaps it is not necessary to be quite so hard on him, particularly if he was doing his work with teaching in mind. To be sure, there are things in Euclid's *Elements* that are 'elementary' in the sense of easy; quite a lot of them, ranging across simple constructions with lines and circles and triangles through the properties of numbers (even times even is even) and those of ratios. But the book also contains a great deal that is eye-wateringly hard by any standard whatever, and which requires a real talent for mathematics, and a lot of time spent on its study. An example is Euclid's exhaustive classification – in book 10 of the *Elements* – of the different ways two lengths can make a ratio that is impossible to express using whole numbers: there are 115 propositions about the subject. Another is the construction of the tetrahedron, cube, octahedron, dodecahedron and icosahedron, and their relative sizes when set inside a sphere that just touches the corners of each of them. It is certainly possible to imagine students for whom Theon's smoothing, explaining and organising would have been of real help.

Influential he certainly was. Although older texts of the *Elements* remained in circulation, Theon's new edition was taken up enthusiastically, and nearly all of the complete manuscripts that have survived to the present contain what seems to be Theon's version. Mediocrity or not, he had more impact on the Euclidean text than anyone else since Euclid.

The sharp-eyed will have noticed a reference to Theon's daughter. Her name was Hypatia, she was born around 355, and early sources suggest her mathematical talents surpassed those of her father, although only the titles of her works survive. These seem to have included commentaries and possibly independent works in mathematics and astronomy.

Theon was living in a changing world. The emperor Constantine had converted – at least formally – to Christianity in 312, and removed the legal obstacles to Christian worship the following year. Churches were opening and – slowly – temples were closing across the Greek world. During Theon's lifetime the temple of Serapis in Alexandria was destroyed: seat of Ptolemy's novel cult, it had been closed since 325, but it was probably one of the main remaining libraries in the city.

The Alexandrian reputation for mob violence continued, and Hypatia herself was one of its victims. By the early 390s she was active as a teacher of philosophy, with an established circle of students. Her teaching included geometry, and she is therefore one of the last recorded individuals who continued the tradition of geometrical teaching in the city of the *Elements*' birth: a well-known figure, even something of a celebrity in the city. Most of her identifiable students were – or later became – Christians, and her paganism does not seem to have been felt as a problem. But in the 410s her supporter the Roman prefect (Orestes, another Christian) came into violent conflict with the Alexandrian patriarch, and in March 415 Hypatia herself was murdered in something like reprisal:

a victim, in the words of one ancient historian, of political jealousy. Since the nineteenth century the tragedy has been often – and often sensationally – fictionalised.

One of Theon's astronomical commentaries bears an ambiguous heading referring to his daughter. It has been variously interpreted as implying that Hypatia checked or revised that part of the commentary (or more), or that she edited or revised part or all of the text on which Theon was commenting. Certainly it seems there was a period when she and Theon worked together on mathematical and astronomical material. But the surviving evidence does not link her with his edition of Euclid's *Elements*. It is possible she did work on it with him, and tempting to speculate that the future teacher of geometry left her mark, somewhere, on the text of the *Elements*. But Theon's career certainly began before his period of collaboration with his daughter, and most probably continued after it, meaning that he could equally well have edited Euclid independently of her. As with so much about the transmission of the *Elements*, it is impossible to be sure.

Stephanos the scribe

Euclid in Byzantium

Constantinople, in the 6,397th year of the world (AD 888). The scriptorium in one of the city's monasteries.

The last page of vellum turns. The scribe flexes his weary fingers, stretches his aching back, and adds a final note:

> Written in the hand of the clerk Stephanos in the month of September, the seventh indiction, the year of the world 6397. Purchased by Arethas of Patras, the price of the book 14 coins.

During the fifth century the Roman imperial system collapsed in the western part of the empire, from Britain to Portugal to the Balkans. By 480 there was no emperor. In the eastern part, which spoke Greek and had had its own separate emperor since 395, the empire flourished for another 1,000 years. The seat of its power was in Byzantium.

Byzántion. Constantinople. Another immense vanity project, it was founded in 324 as a new de facto Eastern capital, and designated the official capital from 330. A university of sorts opened in 425.

This empire was less prosperous, and less literate than its predecessor. Schools – the universities of the period – existed at Alexandria, Antioch, Athens, Beirut, Constantinople and Gaza. But they declined, and by the middle of the sixth century only those at Constantinople and Alexandria remained. The latter was conquered by Persia in 619, ruled by the caliphs from 641. The Greek-speaking world collapsed; Greek culture was now largely that of a single city, Constantinople.

If the barbarians had provided a solution of sorts to the problems of Rome's unwieldy empire, the great contraction of culture provided a similar sort of solution to the unwieldy mass of Greek literature. The production of Greek books slowed dramatically, and what was not salvaged at Constantinople was, by and large, not salvaged at all. The worms got the rest, and the Greek world's library was a manageable size once again.

For teaching purposes, the Byzantine scholars organised matters into a literary cycle and a scientific cycle, together making the 'seven liberal arts' that were also studied in the West. The list of seven subjects went back to classical Athens; it is not really certain when or where it was formalised into a curriculum, perhaps around 100 BC. On the literary side there were grammar, rhetoric and dialectic (logic), which amounted to three different ways of reading the favoured ancient authors. On the mathematical side there were arithmetic, geometry, astronomy and music. The authorities for these subjects ranged from late classical authors such as Aristoxenus on music, down to Greek authors of the Roman period such as Claudius Ptolemy on astronomy. On geometry, of course, the text to study was Euclid's. For the first time, then, it is clear that knowledge of the *Elements* was part of a good education, mastery of it a requisite for those aspiring to join the cultural elite. Or at least, presumably, mastery of the easier parts.

Saving culture with limited resources, particularly when the priority was teaching tomorrow's orators, did odd things to texts. It did them to the *Elements*. Scribes made mistakes; when they were copying mathematics that they did not always understand, they made particular kinds of mistakes, confusing numbers, abbreviated words and the names of geometrical points, all of which used the same set of Greek capital letters, sometimes in the same sentence.

Since readability rather than literality was their main value, scribes also made deliberate improvements, standardising notation, tidying up diagrams, even regularising the logical structure of proofs. Some compared the two versions of the *Elements* now in circulation – Theon's edition and the version that preceded it – and adopted whichever text they preferred at any given point. Coherence and mathematical perfection resulted, perhaps, but at the expense of obscuring Euclid's distinctive voice.

All this was taking place not on the papyrus rolls that Euclid used but on codices – books with pages – that had gradually supplanted them. A single codex could be large enough to hold the whole of the *Elements*, with commentary in the margins too: and that created further possibilities for confusion, when scribes copied pieces of commentary into the main text by mistake, or adopted as main-text readings what were proposed in the commentary as alternatives or improvements. Even when that did not happen, fragments of commentary became enduringly attached to the text as 'scholia', transmitted for generations and often completely detached from their original commentary context.

It was probably in the fifth or sixth century, too, that someone added Hypsicles' treatise on the solids to the *Elements*, calling it book 14, and there it remained, enough scribes accepting the addition to give it a sort of authenticity of its own as a completion of what Euclid had supposedly left incomplete. Most likely in the same period, a fifteenth book was also added, addressing further questions in solid geometry. Although it was sometimes also associated with the name of Hypsicles, it was in reality a combination of material from at least three authors, of whom the latest was Isidorus of Miletus, the sixth-century architect of Hagia Sophia in Constantinople. It may indeed have been one of his pupils who assembled the fifteenth book and added it to the *Elements*.

The Byzantine literary world itself suffered serious disruption from around 550 to 850, with many of its institutions dormant or closed. The empire was at war with the caliphate (almost always) and the Bulgarian khanate (usually) and in conflict with the Roman Pope and the Carolingian empire (occasionally). And it was stirred internally by seemingly endless theological controversy and the hopeless muddle of the imperial succession. In the ninth century came a period of self-conscious revival and rediscovery; the imperial university reopened, and the names of certain teachers have been preserved: Leo the mathematician, Theodore the geometrician, Theodegius the astronomer, Cometas the literary scholar.

These scholars collected books. Leo, for instance, undoubtedly had copies of the works of Euclid; he also had copies of Apollonius' book on conic sections and of works by Theon and by Proclus, as well as those of Archimedes and very likely of the astronomer Ptolemy. Books were now being copied in a new, 'minuscule' handwriting, smaller and easier to read than the old all-capital style, and better supplied with breath-marks and accents. And scribes still perhaps blundered, improved, and sifted the existing versions of texts for what struck them as best. Half the definitions and a third of the propositions in the *Elements* were by now affected by changes from a thousand years of commentary and editing.

Arethas of Patras was another of the scholars of this period, and evidently also a man who loved beautiful books. Born shortly after 850 in western Greece, he came to Constantinople and received a comprehensive classical education, reading widely in ancient literature. His career was in the church; he became a deacon by 895, and in 902 or 903 Archbishop of Caesarea, the most senior see dependent on Constantinople. He was an author and editor, and the writer of a commentary on the *Apocalypse*. A few of his letters survive, and a few of his epigrams. Along the way, he collected books.

He owned, by his death, twenty-four of the dialogues of Plato, and the main works of Aristotle. He had copies of the ancient authors Lucian, Aristides, Dio Chrysostom, Aurelius, and of Christian writers such as Justin, Athenagoras, Clement of Alexandria and the church historian Eusebius. The best of Arethas' books were masterpieces of calligraphy on high-quality parchment. He paid twenty-one gold pieces for his copy of the works of Plato, when Civil Service salaries started at seventy-two per year. He commissioned his books from professional scribes, most of them monks; some were the finest calligraphers of their period. (A measure of the high-pressure perfectionism that ruled in the Byzantine scriptoria is that one of them had a specific penalty for breaking your pen in anger.)

And, of course, Arethas obtained a copy of the *Elements*. In fact, it seems to have been the first book he commissioned; perhaps in connection with the completion of his education in the seven liberal arts. Stephanos, its scribe, was one of the most accomplished of all Byzantine calligraphers. He worked for various patrons; copies survive of the Acts of the Apostles and the New Testament Epistles in his hand, and works by such authors as Ptolemy, Porphyry and Proclus in hands that may be his. He developed a characteristic decorative scheme in blue and gold, with cypresses, columns, lanterns, crosses and geometric figures such as rhombuses, circles, squares and rectangles. And he developed a spacious layout, with wide margins where notes and commentary could appear without making the page look messy.

The *Elements* in 888.

All of this he brought to Arethas' copy of the *Elements* in 888: 388 leaves of parchment in forty-eight gatherings, bound together to make a thick book seven inches by nine. At up to twenty-six lines to a page, there were nearly 20,000 lines of text all told, in a pleasing reddish-brown ink. Stephanos reported at the end of book thirteen that what he had copied was 'Theon's edition'; he went on to copy out books 14 and 15 as 'Hypsicles' addition'.

The diagrams were drawn – it is hard to be certain whether Stephanos did them himself – neatly, with rule and compass, and their labels added in capital letters. They were inserted into spaces in the main text that had been left blank on purpose: always on the right, and always a good legible size.

Arethas didn't like his books just to look well on the shelves; they were not merely prestige objects to be owned for the sake of display. He read them, and annotated their margins in his own hand. In his Aristotle, he annotated a huge amount; in his Plato, little. Often he copied the remarks of others, but he certainly made contributions of his own too, albeit none of great importance (and his attempts at the correction of texts have generally been regretted by later scholars). He explained, commended, criticised, got angry: carried on a dialogue with the authors.

He did a great deal of this in his *Elements*. Evidently he wrote in commentary from other copies of the book, including fragments from ancient geometers and other accretions, some of them illustrated with their own diagrams. But about fifty of his additions appear in no other surviving copy of the *Elements*, and may well be Arethas' own work. He corrected the scribe, added comments, or made brief remarks like 'nice', 'note', or just inserted a tiny ornament. On one page he added a comment about the addition and subtraction of fractions, which originated from a lecture by Leo the geometer. Near the start and end of the volume he wrote two epigrams about Euclid.

Arethas died sometime in the 930s. His was an impressive library but not a huge one; there were others collecting books, and his may well have been unremarkable in its own day. Nor did he always obtain the best texts of the works in which he was interested. And his reputation as a scholar has never at any time stood very high. But his library matters because, in part, it has survived to the present, providing some of the earliest known copies of some famous texts.

Eight of Arethas' books survive, scattered today from Florence to Moscow. His copies of the works of Plato and Aristotle are important witnesses to the texts they contain, and the world owes to Arethas the only surviving copy of the *Meditations* of the emperor Marcus Aurelius, that long-lived favourite of later Stoic thought: Arethas had found an old manuscript, 'not entirely fallen apart', and had it recopied. His *Elements* is one of the two oldest complete copies of the text to survive, the other being an undated ninth-century book also from Byzantium.

It has fared no worse than many another artefact from the period. From about the tenth to the fourteenth century it was evidently much used and annotated. Owners and readers added to Arethas' marginalia, sometimes in large, ugly handwriting and sometimes including large careless diagrams, which did something to spoil the look of the book. Someone added numbers to help understand one of the diagrams. Worse, a few early pages became detached and were lost, so that the first fourteen propositions of book 1 had to be replaced by a copy in a different hand.

The exact whereabouts of the manuscript during all this is uncertain; Greek books were coming to western Europe by various routes from the time of the Crusades onwards, but there is no sign of this particular volume in any records until it turned up in (probably) 1748, acquired for the collection of Jacques Philippe D'Orville, the Dutch-born son of a French family, who travelled in France, Italy and Germany before settling as a professor back in his native Amsterdam.

D'Orville had gradually acquired a large collection of manuscripts, among which Arethas' *Elements* stood out as one of the oldest. On his death, the books passed to his son John and then his grandson; next they were sold to one J. Cleaver Banks, apparently a book dealer, and then in 1804 nearly all of the manuscripts were sold again to the Bodleian Library in Oxford, fetching £1,025. Recatalogued, the *Elements* became manuscript D'Orville 301.

The book remains in the Bodleian. Today, Arethas' *Elements* is the oldest surviving manuscript of an ancient Greek author to bear a date. It is, indeed, the first dated manuscript in Greek minuscule writing, setting aside religious texts. Despite its age and its long journey – indeed, because of them – it is still beautiful. It is hard to look at it and not be moved.

Al-Hajjaj

Euclid in Baghdad

Baghdad, 14 Safar 204 AH (10 August 819). Dawn over the Round City. A little boy stands in a crowd.

> I saw the caliph Ma'mun on his return from Khurasan. He had just left the Iron Gate and was on his way to Rusafa. The people were drawn up in two lines [to watch the caliph and his entourage go by] and my father held me up in his arms and said 'That is Ma'mun and this is the year [two hundred] and four.' I have remembered these words ever since: I was four at the time.

After six years of siege, Abu al-Abbas al-Ma'mun ibn Harun al-Rashid was returning to the city of his fathers. Heir to the caliphate, he was the inheritor of the spectacular conquests of the early Arab armies. Stretching across Egypt, the Fertile Crescent, Persia and India, it was almost the whole of what Alexander had taken a millennium before, and this empire was proving far more durable.

Of the three great heirs to the Roman Empire – the Greeks, the

Arabs, the Latins – it was the Arabs who were the most dynamic, who most rapidly transformed a region beyond recognition. The vast world ruled by the caliphs stretched 4,000 miles from the Atlantic to the Oxus, funnelling goods, people and ideas through – and into – Iraq and Syria. Its wealth was incalculable.

Ma'mun's dynasty, the Abbasids, came to power in 750. It would provide an unbroken line of caliphs for half a millennium, though its apogee came early, in the first century or two. Their values were universal and cosmopolitan, and they created a centralised Arabic, Islamic state around the caliph as commander, leader and lawmaker, with Turks and Iranians, Christians and Zoroastrians assimilated, employed and often powerful in their own right. Their story is the stuff of legend: Harun al-Rashid raiding against the Byzantines; the rise and fall of the Barmakid family; the execution of Ja'far, once the caliph's right-hand man.

The Abbasids moved their capital – after a period of hesitation and experiment – to Baghdad in 762. It was the heartland of their political support and of the sources of their wealth: the rich alluvial plain of southern Iraq, watered by the Tigris and the Euphrates, growing wheat and barley, dates and other fruits.

Baghdad was close to the confluence of the two rivers: a great inland port as well as a centre of overland trade. Soon it became known as the Round City. The central government complex was a walled and ditched circle, built around a palace and a mosque, with wide arcades and generous courts. Suburbs and markets grew up around it; a separate palace – al-Khuld, the Palace of Eternity – was built beside the Tigris.

The city was a monument to the caliph who founded it, Abu Ja'far al-Mansur:

> In the entire world, there has not been a city which could compare with Baghdad in size and splendour, or in the number of scholars and great personalities. The distinction of the notables and general

The palaces of Baghdad.

populace serves to distinguish Baghdad from other cities, as does the vastness of its districts, the extent of its borders, and the great number of residences and palaces. Consider the numerous roads, thoroughfares, and localities, the markets and streets, the lanes, mosques and bathhouses, and the high roads and shops – all of these distinguish this city from all others, as does the pure air, the sweet water, and the cool shade. There is no place which is as temperate in summer and winter, and as salubrious in spring and autumn.

The cultural mix in Iraq was an extraordinary one. There were Aramaic-speaking Christians and Jews in the country, Persian-speakers in the cities, and Arabs both Christian and Muslim. Meanwhile in Syria, Palestine and Egypt, Greek continued to be spoken alongside Arabic, as well as the Syriac and Hebrew into which quantities of Greek writing had now been translated. Along the lengthy border with Byzantium there were not just military

skirmishes but cultural contact, bringing to the caliphate an aware-
ness of just how much ancient Greek learning was still preserved.

Some Persian and Greek books had already been translated into
Arabic, and now that the Abbasids were securely in power and
their capital city was under construction, they turned to learning.

> They had heard some mention of them by the bishops and priests
> among [their] Christian subjects, and man's ability to think has
> (in any case) aspirations in the direction of the intellectual sciences.
> Abu Ja'far al-Mansur, therefore, sent to the Byzantine Emperor
> and asked him to send him translations of mathematical works.
> The Emperor sent him Euclid's book and some works on physics.
> The Muslims read them and studied their contents. Their desire
> to obtain the rest of them grew.

In this first wave of translations are also mentioned works of
Aristotle, Ptolemy and other Greek writers, as well as Persian and
Syriac books and even those of Indian writers on astronomy.

The Arabic translator of Euclid was al-Hajjaj ibn Yusuf ibn
Matar, the son of a scholarly Christian family who was present
at the foundation of Baghdad. Nothing more is recorded of his
personal life or his activities: just that he translated Euclid's
Elements and Claudius Ptolemy's astronomical work the *Almagest*
during the decades around 800. One source places the initial
arrival of a Greek copy of the *Elements* as early as 775. The
Arabic sources report that al-Hajjaj's version of Euclid was made
at the request of al-Mansur's successor Harun al-Rashid; some
say of his vizier.

What was it like, this Arabic *Elements*? Some of the scholars
of this period seem to have translated the sense of their texts
rather than working word for word, but others stayed as close
as possible to their models. The latter is perhaps more likely for
this classic mathematical book; in any event, it seems unlikely

that al-Hajjaj substantially reorganised the Euclidean text. But his was quite a succinct version of the *Elements*: he appears to have worked from a version somewhat shorter than the text that Stephanos copied in Byzantium, lacking a proposition here and there and often missing some of the repetitious setting-out, specification and conclusion to be found in the longer Greek versions. It also quite probably lacked books 14 and 15. The ravages of time have left more than twenty manuscripts of the *Elements* in Arabic, but none seems to contain al-Hajjaj's version in anything much like its pure form, making its detailed character largely a matter of conjecture.

The burgeoning translation movement – and the reception of Euclid's *Elements* in Arabic – were severely disrupted about half a century after the foundation of Baghdad. A war of succession broke out in 196 AH (AD 811) between the brothers al-Amin and al-Ma'mun, sons of Harun al-Rashid, and for more than a year al-Amin was besieged in Baghdad by his enemy's forces. The city was devastated; suburbs were demolished and trenches dug; palaces were dismantled and their materials sold. Al-Amin was murdered in 198/813 but the fighting continued for another six years, with pillage and banditry from soldiers and vigilantes. Cultural and intellectual life surely came to a standstill.

Al-Ma'mun's return to Baghdad after it was all over marked a new beginning. In the midst of pacifying and rebuilding, he and his ruling circle developed a court culture of refinement and sophistication, with a lively interest in the translation movement and the scientific texts it was accumulating. More than ever, Baghdad now became a magnet for texts and ideas: alongside Chang'an in the distant Tang empire it was the most literate city in the world. Al-Ma'mun founded (or perhaps re-founded) a library, the 'House of Wisdom', to store the scientific and philosophical knowledge that was being collected.

Al-Ma'mun's personal enthusiasm was critically important, and the court followed his lead; courtiers used the patronage of translators as a way to establish their own prestige and competed in the hunt for manuscripts. The well-off outside the court, and the military elite, also became patrons of intellectuals. There were new missions to Byzantium, and collections of manuscripts came back to Baghdad for translation. Large personal libraries were amassed, and at one point more than a hundred booksellers worked in Baghdad. At its peak the translation movement was supported by the whole elite of Abbasid society, from princes and courtiers to officials and generals; it cut across ethnic and religious groups, involving speakers of Arabic, Syriac and Persian, Muslims, Christians, Zoroastrians and pagans.

Perhaps the most famous sponsors of translation were the three sons of Musa ibn-Shakir, collectively known as the Banu Musa. Close to al-Ma'mun, they spent lavishly, supporting a circle of professional translators and book-finders with monthly salaries. They took a particular interest in the mathematical sciences, but they, like the rest of Abbasid society, were willing to range across the whole of science and philosophy: astrology, alchemy, metaphysics, ethics, physics, zoology, botany, logic, medicine, pharmacology, military science, falconry . . .

As translation became a profession, the translators' collective knowledge of Greek was improving, and the accuracy and range of their scientific terminology were increasing, meaning that older translations went out of date or needed to be updated. The early Arabic translators of scientific texts were, indeed, faced with an extraordinary task. Arabic is a language quite unrelated to Greek, with entirely different ways of managing syntax and word formation. For many technical terms, new Arabic words had to be devised, by borrowing or modifying Greek (or Syriac) words or by inventing new terms. The translators eventually created a scientific vocabulary that would be used for centuries, from North Africa to China.

In this heady atmosphere al-Hajjaj saw that it would be to his advantage to prepare a new version of his *Elements*. Technical terminology was still unstable and there may have been value in updating his version for that reason; it may be that he had seen new Greek texts that he wished to take into account. What later Arabic historians reported was that he decided to seek the favour of al-Ma'mun (or perhaps was commissioned by his vizier) by correcting, clarifying and shortening his *Elements*, making a version for specialists in which anything superfluous was deleted and anything missing was added.

Al-Hajjaj, or if not he then someone else in this first period of the Arabic *Elements*, also started a habit of giving names to certain of its propositions. One was called 'the Mamunian' – *al-ma'muni* – after the caliph who was al-Hajjaj's second patron. Another – the so-called Pythagoras' theorem – was called 'the one with two horns', applying rather incongruously to its diagram a traditional Arabic surname for Alexander the Great. Later in the book came the 'goose's foot', the 'peacock's tail' and even the 'devil'. Presumably the names made it easier to remember which proposition was which or to recall the general shape of some of the diagrams: foot, tail or horns.

The mere fact that there was an audience of specialists – trained, interested people – to read this new, leaner version of the Arabic *Elements* is an indication of what the translation movement was achieving. Over the course of two centuries at least eighty Greek authors would be translated, including the big names such as Aristotle and Plato. Almost everything of Greek science and philosophy that had survived the wreck of the Roman Empire was rendered into Arabic during this period. It was a renaissance of learning on the largest possible scale, with profound consequences for the formation of an Islamic intellectual culture, which assimilated and developed not just Greek but Indian and Persian learning too.

The continuing flow of manuscripts to Baghdad meant that the story of Euclid in Arabic did not end with al-Hajjaj. Later in the ninth century another translator produced a fresh version. It was afterwards revised by a third man: Thabit ibn Qurra, one of the Banu Musa's protégés. The *Elements* was becoming part of the stock of Arabic mathematical learning, to be worked over and rethought as needed.

It had become, indeed, the most important mathematical text in the Islamic world, central to the teaching of geometry and a stimulus to all manner of new mathematical work. It was corrected, summarised, added to, excerpted and frequently commented upon. Perhaps fifty Arabic commentaries on the *Elements* survive today, as well as Arabic translations of ancient Greek commentaries. Omar Khayyam, better known in the West as a poet, wrote a treatise explaining the difficulties in Euclid's postulates; Ibn Sina, called Avicenna in the medieval Latin world, made a summary of the *Elements* as the geometrical part of an encyclopaedic work. In the tenth century an early writer on fractions went by the nickname 'al-Uqlidisi': meaning, most likely, that he made his living copying manuscripts of Euclid (Uqlidis). The Round City of al-Mansur was a ruin by then, but the Arabic *Elements* to which it had given birth lived on.

Adelard

The Latin Euclid

Bath, in south-west England, early in the twelfth century. Two men and two books. One book is in Arabic, and one of the men reads from it, translating as he goes into – probably – Spanish. The other listens to the Spanish and writes it down in Latin in his book. Word after word, page after page. There are stumbles over vocabulary, discussions about terms. The Latin Euclid is, gradually, born.

In the medieval story of Euclid, the Arabic versions hold unparalleled importance. From them, the text was translated on into versions in Sanskrit, Persian, perhaps Syriac, and other languages, some of which survive only in fragments. And it was translated into Latin.

For around half a millennium, educated citizens of the Western Roman Empire read Greek geometry in Greek, if they read it at all. Then for half a millennium and more after the Western Empire collapsed, they were reduced to mostly proof-free summaries of Euclidean results in Latin (see this book's chapter on Hyginus in Part III for a fuller story). It is not clear that the whole – or anything like the whole – of the *Elements* was translated into Latin before the year 1000.

But, from late in the eleventh century, the Christians reconquered Spain and the Normans dominated Sicily, creating new possibilities for books in Greek or Arabic to pass into the Latin-speaking world. The new situation also made it much more possible for people who could read and write Latin to come into contact with those who could read Greek or Arabic. (The contemporaneous crusades in the eastern Mediterranean perhaps did rather less for cultural exchange, although the sale or plunder of Islamic libraries was certainly a factor in the movement of texts in this period. Saladin auctioned the famous Fatimid library from AD 1171 onwards, for instance; and in 1204 Constantinople itself, summit of Byzantine manuscript culture, was taken by the Latins.)

What followed was a translation movement comparable in its overall scope to the translation movement at Baghdad two centuries earlier: but this was decentralised, driven not by court patronage so much as by a new breed of sometimes itinerant lay teachers, hungry for texts. Translators worked at Barcelona, Tarazona, Segovia, León, Pamplona, and also north of the Pyrenees in Toulouse, Béziers, Narbonne and Marseilles. An intense enthusiasm for translation into Latin would last at least a century. Toledo would become a particularly important centre for translation and scholarship. Much of Arabic science and learning was translated into Latin during the twelfth century, and the consequences were incalculable.

Naturally the Arabic *Elements* was a part of this. The book was translated into Latin several times during the twelfth century, by scholars whose origins ranged from Chester in northern England to Cremona in Lombardy. The very first was done by a man whose life would qualify as an adventurous one in almost any period: Adelard of Bath.

Fastred was one of the tenants of the Bishop of Wells; he held land at Wells, Yatton and Banwell, all in Somerset. His son Adelard

– Athelardus – was born in about 1080: less than a generation after the Norman conquest of England. He was clever, perhaps extremely so, and the combination of his family's means and perhaps patronage, and the new situation of a Norman-controlled England, combined to provide exciting possibilities for his education. He went to France, to Tours, studied the liberal arts and showed a particular interest in astronomy. He learned music and played – by invitation – before the queen. He wrote a dramatic dialogue in the manner of Boethius, exhorting youths to the study of philosophy.

Adelard of Bath.

He could have become a career cleric or perhaps an administrator at court. But instead he went travelling. Leaving behind a nephew to concentrate on the tamer Gallic studies (the nephew, from Adelard's own writings about his life, may possibly be a fiction), he set out in search of the novelty of the age: Arabic learning.

In an instance of what might well be called Crusade tourism, he made his way to the Norman kingdom of Antioch in Syria, taking in Salerno and perhaps Sicily and Cilicia along the way. He saw an earthquake in Syria, a pneumatic experiment in Italy, met a Greek philosopher and learned that light travels faster than sound. Adelard, in other words, appears to have travelled from England to the border with the Islamic empire – without any official protection or support – just to seek the knowledge of that other civilisation. The success of the first Crusade in 1098 had produced something of a vogue for visiting the Middle East, but it was far from usual for scholars to travel in that direction for the specific purpose of encountering Arabic science.

Adelard certainly succeeded in encountering Arabic scientific texts, including astrological books, astronomical tables and Euclid's *Elements*. But the details become lamentably obscure at this point. His astronomical tables were calculated for use at Cordova, and the most natural place to find an Arabic *Elements* would also have been reconquered Spain. But Adelard, despite a tendency to show off about his travels, never said that he had been to the Iberian Peninsula. Thus, there is something of a mystery.

During the 1120s, Adelard returned to England. He turns up in legal and court records in the area of Bath, though his name was not quite unusual enough to be sure that every such report refers to the same man. This was a turbulent period for England, and it looks as though Adelard may have supported first King Stephen and later his rival Matilda; he dedicated a book to her son, the future Henry II. He was quite probably associated with a circle of learned clerics around the Bishop of Hereford and Walcher and Prior of Malvern. He wrote on astrology, translated the astronomical tables of al-Khwarizmi (head of the library of the House of Wisdom in Baghdad) and wrote a work of natural philosophy drawing in a general way on his knowledge of Arabic ideas. He taught, and wrote a little book about falconry. And he translated the *Elements* of Euclid from Arabic, probably shortly before 1130.

As one historian remarks, 'nothing in his life so far has prepared us for this'. It is not altogether clear what had prepared Adelard for it, who nowhere stated that he could read Arabic. The writings of his earlier life betray no special knowledge of geometry nor particular interest in the subject, although he certainly knew some practical writings about surveying and may have seen a Latin summary of the *Elements*.

Prepared or not, it is remarkable to think of translating a book of the length and complexity of the *Elements* in the conditions of the time: the more so as Adelard presumably did the work after he was back in Bath, the better part of 1,000 miles from any location where Arabic was spoken. Yet it is likely that he did have some linguistic help.

Later reports show that Latin scholars in centres such as Toledo employed the help of local experts: there would be a text in Arabic, a local person who spoke both Arabic and Spanish, and a visiting or immigrant scholar who could understand Spanish and write Latin. Translation would be effectively a two-stage process, with the text passing through a spoken Spanish version along the way, and much of the intellectual work of understanding the text done by the arabophone 'assistant' rather than the Latin 'scholar'.

It is quite conceivable that Adelard had an assistant, now lost to history, or something of this kind: or perhaps someone who helped him to render the Arabic directly into Latin. Indeed, it seems less than plausible that he made his way through the conceptual and terminological complexity of the Arabic *Elements* without some such help. At one stage, the learned circle in Hereford and Malvern had included Petrus Alfonsi from Huesca in Aragon, a converted Spanish Jew who became physician to King Henry. He worked on astronomical material, and may well have provided the base translation from which Adelard produced his version of the tables of al-Khwarizmi. Petrus moved to France too early, it seems, to have worked on Adelard's *Elements*, but his example indicates the kind of collaboration that might have been involved, and the kind of person who may have stood, uncredited, behind Adelard's work.

Be that as it may, the Latin *Elements* was a remarkable production. It clearly showed the linguistic stages it had been through, and the struggles of this first medieval translator to find the right Latin words for what he needed to say. In one place Adelard tried three different phrases in succession, all trying to express the idea of 'a ratio repeated three times'. Elsewhere he hesitated between words meaning 'difference' or 'remainder' when talking about the result of taking a quantity away from another.

It is plain to see that he was not working from the Greek text. He used none of the Greek loan-words that would later become common in Latin, such as *hypotenusa, parallelogrammum, ysosceles*. Instead his Latin contained a wealth of terms that were simply Arabic: *alkamud* (perpendicular), *alkaida* (base), *mutekefia* (proportional), *elmugecem* (solid), *elmugmez* (pentagon). Even when he knew the equivalent Latin words, Adelard sometimes deliberately adopted an Arabic terminology: *elkora* for sphere and *elmukaab* for cube, for instance.

Also from Adelard's Arabic source came the names for some of the propositions, of which he made some strange gibberish: *elefuga, dulcarnon, thenep atoz, seqqlebiz*. Some of them nevertheless stuck, notably *dulcarnon*, which went on to be mentioned by Chaucer and would later facilitate the popular pun 'dull carnon'. He also took from the Arabic *Elements* a curious habit of introducing the different parts of each proposition with fixed, formal phrases: *Now we must prove . . . So I say that . . . For the sake of argument . . . The reason why.* These do not appear in any surviving Greek version, and were surely not invented by Adelard; they provide a glimpse of features of his Arabic source which are otherwise lost.

Of the two main Arabic versions of the *Elements*, the modern consensus is that Adelard's version seems to be derived from the earlier, that of al-Hajjaj. The clearest piece of evidence is that the number of propositions in Adelard's Euclid agrees with al-Hajjaj in ten of the fifteen books, whereas in four of those books the later Arabic version, by Thabit, has a different number of propositions. Similarities of wording support the conclusion, but several matters complicate it. Al-Hajjaj's version is largely lost; Adelard's survives only incompletely (book 9 and part of book 10 have been lost). No surviving Arabic manuscript is close enough to Adelard's version to have been his actual source, and it is possible that what he worked from was not a 'pure' al-Hajjaj text but one with some modifications. For further intrigue, a fifteenth- or sixteenth-century manuscript containing much of book 1 in Syriac agrees almost word for word with Adelard; even the lettering on the diagrams matches. Does this mean that the Syriac translator and Adelard used the same, Arabic source, now lost? Certainly the last word has not yet been written on the subject.

Adelard's was a remarkable achievement, and his name became celebrated, attached to Euclidean material by later scribes whether it was really his or not. His opinions on some details in the arithmetic of Boethius were carefully recorded by a student; one 'Ocreatus' dedicated an arithmetical work to him. His version of the *Elements* was a key moment when a mathematical tradition from the shores of the Mediterranean acquired a presence in northern Europe, from there to pass – a few generations later – into the chain of new universities stretching from Italy to England.

By about 1150 Adelard was dead, but the translation work of the twelfth-century scholars went on, and Latin attention to the

Elements of Euclid gathered pace. The translators of the period worked with evident excitement on a mass of material to which their access was somewhat haphazard, and they worked in some isolation from one another. Texts were translated more than once, and tangles of different versions emerged as new Arabic texts were obtained or new ways were tried of amalgamating them. For the *Elements* alone, the cast of characters becomes quite large and the names of the translators and editors give a sense of the range of people interested in the book: Hermann of Carinthia, Robert of Chester, Gerard of Cremona, Ioannes of Tinemue. Gerard, in the mid-twelfth century, actually moved to Toledo, becoming the century's most prolific translator from Arabic: he rendered Euclidean commentaries and Euclid's *Data* as well as the *Elements*. In Sicily, where there was a tradition of bilingualism and trilingualism, someone translated the *Elements* directly from Greek into Latin, late in the twelfth century. A number of other reworkings have not yet been studied in full even today: some of them exist only in single manuscripts and represent complex mixtures of different versions.

Thus all of the early versions, including Adelard's own, came to be eclipsed. First by the version of Robert of Chester, popular throughout the twelfth century and, ironically, often attributed in the manuscripts to Adelard himself (Robert may have been a student of Adelard, which does something to explain the confusion). And, second, by the work of Campanus of Novara. Writing in the 1250s, he put together what would become *the* Latin Euclid of the later Middle Ages. He used Robert's version for the statements of each proposition: material that – much of it – went back to Adelard. But he composed new versions of the proofs, bringing together material from other versions and commentaries. And he added to the book, with new propositions and definitions from various sources: the impulse to modify and improve the *Elements* was proving as durable as the book itself.

A hundred manuscripts of Campanus' *Elements* survive today: it was a bestseller in its day. For Adelard's version there are just four, none of them complete. But through Campanus, much of what Adelard had done in rendering Euclid's words into Latin survived and flourished for another three centuries, and passed on into the era of print.

Erhard Ratdolt

Printing the *Elements*

May 1482, in La Serenissima: the Republic of Venice. The print shop of Erhard Ratdolt.

In one room, compositors perch on high stools, slotting tiny pieces of lead type into little frames, called composing sticks. A stick holds a few lines of text (upside down and backwards), and when it is full the assembled letters are lifted out in a block, and added into the tray – the forme – in which a whole sheet of text is being assembled. A good compositor can set a few thousand characters – several hundred words – in an hour.

Once a full sheet is made up, the letters (and any pictures) are clamped together and carried into the next room, to the press itself. Out slides the bed from under the press, and the forme full of letters is placed on it and spread with ink. Two frames fold out to hold the paper; they fold back and the whole thing – paper, ink, letters – slides back under the press. Two pressmen pull a long lever to screw the press down, pushing the paper against the inked type and transferring ink onto the paper. They release the lever, slide out and unfold the frames and take out the paper.

The printer's boy (or 'devil') takes it away. Another sheet of paper follows, and another: hour after hour.

The workshop is busy, noisy and dangerous. The type is made of lead; the ink probably contains vitriol, and its manufacture involves burning pitch and boiling oil. The press itself is heavy and as powerful as a wine press, its close ancestor.

From one workshop in the first months of 1482 there come more copies of Euclid's *Elements* than Europe has made in 1,000 years.

The owner and overseer of that workshop, Erhard Ratdolt, was part of the German diaspora that took Gutenberg's invention across Europe in the second half of the fifteenth century. One printing press in the world used movable type in the 1450s; by 1470 such machines existed in fourteen cities, and by 1480 in over a hundred. Born in about 1447, Ratdolt grew up in Augsburg in southern Germany, and after an apprenticeship crossed the Alps by the Brenner Pass, setting up as a printer in the first and finest city he reached: La Serenissima.

Venice in the 1470s.

A commercial and business hub with a fine natural harbour, Venice was at the height of its power and influence in the late fifteenth century: and of its splendour. In the previous half-century many of its famous landmarks had been erected: the facade of the Doge's Palace, the Ca' d'Oro, San Giovanni e Paolo, the Porta della Carta and the Ca' Foscari. Giovanni Mocenigo occupied the doge's throne. A French ambassador described the city:

> I was conducted through the longest street, which they call the Grand Canal, and it is so wide that galleys frequently cross one another; indeed I have seen vessels of four hundred tons or more lie at anchor just by the houses. It is the fairest and best-built street, I think, in the world, and goes quite through the city. The houses are very large and lofty, and built of stone; the old ones are all painted; those of about a hundred years' standing are faced with white marble from Istria, about one hundred miles off, and inlaid with porphyry and serpentine. Within they have, most of them, at least two chambers with gilt ceilings, rich chimney pieces, bedsteads of gold colour, their portals of the same, and exceedingly well furnished. In short, it is the most glorious city that I have ever seen, the most respectful to all ambassadors and strangers, governed with the greatest wisdom, and serving God with the most solemnity.

By 1476 Ratdolt was printing as a member of a three-man partnership, part of an immense expansion of print in the city: from nothing in 1469 to 150 print shops by the end of the century. The combination of convenient sources of paper in northern Italy and generous provision for the protection of texts and technical improvements made the city one of the centres for the new technology. Ratdolt and his partners printed calendars, histories and geographies, a total of eleven books in three years. They were some of the finest printed books ever produced, with ornamental title pages, careful woodcut illustrations, page borders and initial letters, and two-colour texts printed in black and red. Illustrations of eclipses in the calendars had colouring in yellow added by hand. The type, like that of other printers in Venice, was modelled on the handwriting of contemporary scholars; it is admired to this day for its harmony and elegance.

In 1478 the plague threatened to end it all. It afflicted Venice for four years, and at its height 1,500 people were dying every day. Many fled; half the printing houses closed, and Ratdolt's partnership never printed again, although all three of its members survived. Ratdolt's daughter Anna was born in June 1479 when the disaster was at its height.

By the following year he was beginning to rebuild his printing business, this time as sole owner and employer of a full team of compositors, pressmen, engravers, proofreaders, apprentices and assistants. He worked, not surprisingly in the circumstances, with a furious energy. In his first year on his own he produced eight books, and he took pains to reach a broad market, ranging across ecclesiastical works, histories and – a growing specialism of his – mathematical texts.

There had been plans to print the *Elements* before: the astronomer and printer Regiomontanus, under whom Ratdolt had served as apprentice, had proposed an edition in the 1470s but never carried it out. It fell to Ratdolt, then: and at his hands, ten years before Columbus crossed the Atlantic, Euclid made the momentous transition to print.

It was a vast project. Ratdolt chose the large 'folio' format, producing a book about eight inches by seven, and even so it filled 276 pages; nearly seventy sheets of paper had to go under the press, twice each, to produce each copy (each was then folded, so that when bound into the book it formed four pages). The compositors must have set well over half a million characters of type in all.

There were some problems no one would have expected. So many of Euclid's propositions began with 'If . . .' ('*Si* . . .') or 'Let there be . . .' ('*Sit* . . .') that some pages needed up to thirteen decorated S's: the set of decorated initials Ratdolt was using ran out and had to be supplemented by another set that did not match. In a similar way, the incessant reference to points in diagrams labelled A, B, C, D and so on, put pressure on the supply even of normal-sized capital letters. Four different sets of type had to be used in order to complete the book.

Technical challenges apart, Ratdolt took every care to make sure the book was of the highest quality. He seems to have obtained quite a good manuscript of the *Elements*, in Campanus' Latin version: probably part of a celebrated ecclesiastical library deriving from the mid-century papal court, where there was a tradition of well-illustrated mathematical books. The decorated first page was put under the press three times: once for the text, once for the border and once for the title, which was printed in red. Pinholes in the paper were used to line up the separate layers of printing. The type was neat and strikingly accurate, and the decorated border – reused from one of Ratdolt's earlier books – was a beautiful piece of design. The page layout was broadly similar to that of Gutenberg's Bibles, with broad margins modelled ultimately on the layouts of fine medieval manuscripts.

And the diagrams? This was the first really geometrical book to be printed; the first for which a large number of accurate diagrams were needed. Printing diagrams was known to be hard; editions of Ptolemy and Vitruvius were appearing from other presses without diagrams, because they were too difficult to produce. And even in the world of manuscript, it had become common for diagrams to be poorly copied or even omitted because of the time and skill involved in their construction. The *Elements* needed 500 or more, and, as Ratdolt himself pointed out, nothing in geometry could be properly understood without them.

He was determined to get this right, and indeed his decision to print Euclid's *Elements* seems to have depended on a breakthrough he achieved in the technique of printing geometrical diagrams. The usual way to insert illustrations in printed books was to carve them in relief in wood, clamping the block of wood into the frame with the type, inking it and printing from it in just the same way as the text. Good effects could be achieved, but there were limitations: particularly in creating fine lines of even thickness, keeping them straight when they were meant to be straight and circular when they were meant to be circular. It was particularly hard to make it work for geometrical diagrams.

Ratdolt boasted in a preface to his *Elements* that he had a special technique for printing the diagrams, but scholars do not agree about what it was. One suggestion is that he took strips of metal – zinc or copper most likely – bent them into the shapes required, and mounted them in plaster to hold them in place, together with pieces of type that would provide the labels – A, B, C, and so on – within the diagram. Another possibility is that he used cast pieces of metal clamped together much like the pieces of type that made up the words on the page. Or, again, he may have had special tools and techniques for making woodcuts of exceptionally fine quality.

In any case, the wooden or metal diagram could then be clamped into the set-up page of type – Ratdolt invariably placed the diagrams in the margin, which would simplify this part of the process – and its protruding edges inked and printed from just like the rest of the page. However it was done, Ratdolt achieved results of the highest quality, with fine, even lines that were curved or straight as required. He has deservedly been called the most inventive printer after Gutenberg. The result was a book of which Ratdolt could justly be proud, and which increased still further his already high reputation among Venetian printers. No surviving medieval manuscript of the *Elements*, in any language, has diagrams of a similar number and quality.

After a large number of copies had been printed, Ratdolt stopped his press and reset the first nine leaves of the book. He wanted to rearrange some of the diagrams, insert some new ones and generally improve the completeness and clarity of the pages. Such an obsessive care, a commitment to perfection, almost defies belief, even in the proud, skilled world of the early printers.

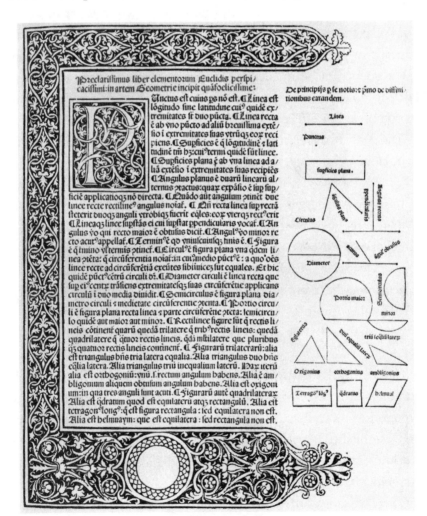

The *Elements* in 1482.

It was a work fit for a king, and Ratdolt added to his book a separately printed leaf dedicating it to the doge. There's no evidence that the doge actually commissioned or paid for the printing of the book (or wanted it), but Ratdolt's dedicatory letter placed it unambiguously under his ruler's protection, casting Giovanni Mocenigo in the role of patron and himself, Ratdolt, in the role of a creator offering up his work. It was a way to boost his own status – higher than a craftsman's, and nearly as high as an author's – as well as an opportunity to display to the world what he could do. For a few presentation copies, he printed on vellum instead of paper, and put gold dust into the ink instead of the usual lamp-black, so as to print the dedicatory letter in flamboyant gold letters. For the doge's own copy, the first page was illuminated by hand to complete the effect.

The dedicatory letter itself spared no pains to emphasise Ratdolt's own cleverness:

> I used to wonder why it is that in your powerful and famous city there are many works of ancient and modern authors being published, but none or few and of little importance are books of mathematics . . . Until now, no one has found a way to make the geometric diagrams . . . I applied myself and with great effort I made the figures, so that the geometric figures are printed with the same ease as the verbal parts of the *Elements*.

Ratdolt probably printed over 1,000 copies of his remarkable edition of the *Elements*. He had set standards for mathematical typography as well as for the printing of geometrical diagrams; he had established a visual language for geometry within the new world of print, similar to the visual language of manuscript geometry but in subtle ways different.

There is sadly little information about where the copies went. A de luxe book is not necessarily a book that is read, although in the intellectual climate of fifteenth-century Italy it is reasonable to suppose that this, the first printed version of such a widely known ancient text, received its fair share of attention. And indeed, the surviving copies show that many a reader responded to the diagrams by working with pen or pencil in hand, re-copying them as the anonymous reader at Elephantine had done and using the blank space in Ratdolt's generous margins as rough paper: continuing the ancient tradition of understanding and learning the ideas the words and diagrams embodied.

Before the century was out, there were a handful more editions of the *Elements* in a Latin abridgement, as well as a pirated reprint of Ratdolt's edition that appeared at Vicenza in 1491. But the *Elements* really took flight during the next century, as print spread and cheapened and printed books decisively took over from manuscripts in many (though not all) situations. About forty substantially different versions of the *Elements* were produced between 1500 and 1600: some were printed several times; thus there was a new printing something like once a year on average. Most were in Latin, but Simon Grynäus edited the Greek text for publication in Basel in 1533, using manuscripts in Venice, Paris and Oxford. The first complete vernacular version to see print was the Italian translation by mathematician Niccolò Tartaglia, printed in 1543. By the end of the century the *Elements* would also be printed in German, French, Spanish, English and Arabic. It is not clear that any text except the Bible was being set up in type more frequently; that pattern would last another two centuries and more.

Ratdolt himself returned to his native Augsburg after ten brilliant years in Venice, there to print for another four decades. He developed a focus on liturgical books, though he also continued to work on scientific texts and histories. He didn't use his novel system for printing geometrical diagrams again, and although he reprinted a number of his Venetian books, the *Elements* was not one of them.

Erhard Ratdolt died, wealthy and highly respected, in March 1528, at about the age of eighty-one. A tribute written by a fellow Venetian printer in 1482 – the year of his *Elements* – might stand as his epitaph:

> The seven arts, abilities granted by the divine power, are amply bestowed on this German from Augsburg, Erhard Ratdolt, who is without peer as a master at composing type and printing books. May he enjoy fame, ever with the favour of the Sister Fates. Many satisfied readers can confirm this wish.

Marget Seymer her hand

Owning the *Elements*

A handwritten inscription in a 1543 printed edition of Euclid's *Elements*, now in the National Library of Wales: 'Marget Seymer her hand'.

Putting the *Elements* into print sent it into the wild, making it common property in a way it had never been before. Printed editions varied very widely, and reflected not just the range of medieval versions of the text but also the new agendas of the Renaissance world, and the commercial imperatives of its printers and publishers.

On the one hand there was Euclid the great author, to be edited and translated with reverence and respect. To be involved with this tradition as editor or commentator was to display one's credentials as a scholar, and the result tended to be large, heavy, expensive books meant for private or institutional libraries, well supplied with both space and money.

But there were also versions of the *Elements* for smaller

pockets. Students at many universities were required to read specifically the first six books of the text, plus sometimes books 11 and 12: that is, the two-dimensional geometry, the theory of ratios, and the more basic three-dimensional geometry. The arithmetic or theory of numbers that made up books 7–9 was left out; so was the very tricky ratio theory in book 10 and the fairly tricky discussion of regular solids in books 13–15. There were, inevitably, printed editions catering to this particular need, containing only the prescribed eight books and usually in smaller formats, meant to be easier to carry around and cheaper to print and purchase.

There were slimmer Euclids still. Scholar and teacher Petrus Ramus in Paris did much to disseminate the belief, also aired by other early printers and editors, that the proofs in the *Elements* were not written by Euclid. Since it was quite true that the proofs in many printed Latin versions went back no earlier than Campanus in the thirteenth century, and since there was no direct evidence one way or the other about the authorship of the proofs to be found in Greek manuscripts of the *Elements*, the charge stuck easily, and Ramus and others were able to find a market for editions of the *Elements* containing just the statement of each geometrical proposition, leaving the provision of proofs and constructions to the preference of the individual teacher or the initiative of the individual reader: geometry on the do-it-yourself model.

Another world again were the editions of the *Elements* in European vernaculars that started to appear in the mid-sixteenth century. Some at least of these seem to have been aimed at the upwardly mobile merchant, anxious to demonstrate cultural sophistication but not anxious to claim a – perhaps non-existent – competence in Greek or Latin. That meant large formats, heavy price tags, copious annotation and the inclusion of learned prefaces and appendices.

Henry Billingsley's English translation of 1570 followed this pattern: it was based on the printed Greek text but included comments and remarks selected from several Latin editions: 'mani-folde additions, Scholies, Annotations and Inventions . . . gathered out of the most famous and chiefe Mathematiciens, both of old time and in our age'. It contained a lengthy preface from the pen of Elizabethan mathematical celebrity John Dee, his remarks at various points in the text and a fold-out chart listing the different branches of mathematics and their relationships. The whole thing filled 1,000 pages, as against the forty-five of Ramus' minimal *Elements*.

So, in the world of print, Euclid's *Elements* was rapidly ceasing to be a single text, and becoming a wide tradition of different texts meant to be used in different ways by different kinds of people. It is immensely tempting to ask how they were actually used: who bought these books and what they really did with the Euclidean texts once they had them in their hands.

At the simplest level the answer is easy: they wrote all over them. There seems to have been a very long-lived convention that readers should write in mathematical books; by the later Middle Ages, owners were even writing on Stephanos' priceless manuscript of the *Elements*. Print offered equally wide margins and less sense of defacing something unique; and in the world of print there was no tendency for marginal annotations to find their way by accident into subsequent copies of the text. Schools and universities seem to have positively encouraged the reading of mathematics with pen in hand, and the result is that up to three-quarters of surviving printed mathematics books from the sixteenth and seventeenth centuries bear readers' annotations.

As a result, quite a lot of information is preserved about how people read Euclid. They read selectively, choosing a few proposi-tions here and there, perhaps marking up the page to show which parts they had worked through or marking up the contents page or the index to the same effect. Surely teachers directed some of this, but surely, too, some of it was idiosyncratic: and there seems to be little evidence that the selections themselves were ever passed from person to person. For example: two surviving copies of the same late seventeenth-century edition of the *Elements* were marked up for use in English universities with selections from the first six books; yet the two sets of selections coincide hardly more than can be accounted for by chance.

Within the portions they chose to study, readers negotiated, sometimes aggressively, with the printed text, about what it should say. They corrected, added and emended. They noticed – and fixed – wrong labels in diagrams or missing lines in proofs. Some authors and editors frankly invited this kind of thing, by adding to the book at the last minute a list of 'errata' they had noticed while it was being printed: but most readers went far beyond any such authorised list in their corrections. Some added whole new propositions, cross-references to other books or portions from ancient or modern commentaries. Some collated more than one edition of the *Elements*, copying sections or details from one into another.

Finally, some readers used the margins of printed books simply to practise their own mathematics: writing out calculations, copying diagrams or repeating proofs until they were satisfied with their own performance. This kind of thing can give a most vivid impression: of learning mathematics as a sort of rehearsal; of mathematics itself as a special type of performance in the Renaissance world, just as it had been in antiquity.

But who were these readers, correctors, selectors and performers? In very many cases it is impossible to know. Some owners did sign their books, but it is not always possible to be sure that the hand of the signature was also responsible for the more interesting annotations found elsewhere in a book. And even when it is, most of the names are mere names, hardly capable of being located in time, place or society.

One such is the Marget Seymer who signed a copy, printed in Paris in 1543, of the *Elementale geometricum* of Johannes Voegelin; a fairly brief compendium drawn from Euclid that enjoyed a vogue in the sixteenth century. Women engaging with Euclid are elusive in this as in most periods, and it would be valuable to know more about her. But she seems to be virtually lost to history. She might be the Marget Seymer, daughter of Robert Seymer, who married Jerom Atwod at All Hallows, Honey Lane, in London on 11 May 1553 and was buried forty years later, in 1593. But she might as easily not be.

The same is true of many another Euclidean owner. Some can be connected with particular universities, schools or colleges; some signatures include a note of the date. Some readers provided other clues, such as by pasting in a grand bookplate with a coat of arms; and of course sometimes the book itself remains in situ in the library of a stately home or a university or college. More often, though, the books have moved, the records are lost and the trail is cold. When books were bought second-hand, some readers unhelpfully obliterated the signatures of previous owners, that might otherwise have helped to establish a trajectory. And the chronology is not always what it seems. One Sandie Hume remarked that Euclid's *Elements* was 'a Very Queen of a Buck', and June Amelia Hume, presumably a relative, also signed and dated the copy. The book was printed in 1719, but she signed it more than a century later, in 1827.

Euclid in Renaissance Europe was thoroughly established as part of culture. It was more than ever – as it had been in the ancient and the medieval worlds – a site for annotation and engagement, for garlands of text draped around Euclid's words themselves. But the very range of its travels makes some of those marks more tantalising than otherwise.

Edward Bernard
Minerva in Oxford

On the endless shelves of the Bodleian Library in Oxford there lives perhaps the most unusual copy of the *Elements* in the world. It has been put together from the separate sheets of four different printed versions of the text, in three different languages: Greek, Arabic and Latin. Its pages are dense with annotations made in the last two decades of the seventeenth century, in the hands of the then professors of geometry and astronomy at Oxford. Textual difficulties are wrestled with, mathematical problems are explained. Algebraic equivalents are provided for propositions about geometry or about ratios.

This polyglot Euclid is bound in two volumes, and it rarely sees the light of day. Its complex annotations are hard to interpret and often simply hard to read. But it is a monument to the kind of attention the text of Euclid's *Elements* was beginning to attract by the years leading up to 1700.

The number of different printings of the *Elements* during the sixteenth and seventeenth centuries was staggering, approaching 300 by the end of that period. Even when there were scores of versions already available, printers and publishers showed no diminishment in their willingness to put the Euclidean text into print yet again: translated, rearranged, simplified, truncated or garbled as it might be. A new cohort of students every year did something to refresh the market, as did publishers' invention of new markets by translating Euclid into modern vernaculars and by cutting, rearranging, annotating and re-presenting the book to appeal to new parts of the social scale.

By the second half of the seventeenth century an enthusiast with money to spend could amass quite a collection of Euclids. Robert Hooke, who, as well as being a founder member of the Royal Society of London and a celebrated experimenter and natural philosopher, was the lecturer on geometry at Gresham College in London, owned thirty-one different editions of the *Elements*, assembled at least in part for the sheer pleasure of collecting.

Yet for all the attention the *Elements* had received from printers and publishers over the two centuries since Ratdolt, it had remained strangely untouched by the kind of textual scholarship that other ancient books received. Version after version was put into print with scant regard for textual accuracy; the surviving Greek manuscripts were consulted only rarely, and the complete Greek text was put into print just once during 200 years (in 1533) – and that, badly.

An institution with a particular interest in ancient mathematical texts was the pair of mathematical professorships set up at Oxford in 1619 by Sir Henry Savile. The Savilian professors, who had their own rooms and a special collection of books within the Bodleian Library in central Oxford, amassed an impressive set of versions of the *Elements*. Savile himself donated not just printed books but manuscripts, and by the mid-seventeenth century Oxford was one of the few cities on earth to possess manuscripts of the *Elements* in Greek, Arabic and Latin.

Savile's statutes for his two professorships aimed to promote a restoration of mathematical learning. They also aimed to refute – or facilitate the refutation of – some of the nonsense he felt was being talked on the European continent about mathematics and its ancient authors. Savile – who once described himself as passionately inflamed with love for geometry – felt a reverence for the Euclidean text that was worlds away from the quick-and-dirty printings it was receiving at the hands of some scholars, and he took particular exception to the notion that the book needed to be pruned or rearranged to make it suitable for the modern student. Polemicising against one cut-down version, he wrote that the *Elements* constituted a 'perfect body' of geometry: pure, shining and virtuous, fit to be admired, studied, and indeed loved. He wished his professors to undertake not just the teaching of the ancient mathematical authors but the study, editing and publication of their texts.

Good intentions lay dormant for a while, partly because in the mid-seventeenth century England had its mind on civil war rather than the editing of ancient mathematics. But late in the century Savile's intentions began to be realised, with new, learned editions of certain texts: works on astronomy, on the mathematics of music and the like, newly edited from the rich manuscripts in Oxford's libraries.

Into this situation came Edward Bernard: born in 1638 in the English Midlands, the elder child of a clergyman, and educated in London at the Merchant Taylors' School and at St John's College, Oxford. He learned the classical languages as well as Hebrew, and evidently had a talent in that direction; by early middle age he also knew some Arabic, Syriac and Coptic. He continued at St John's College as a fellow and, later, a university proctor and college bursar.

Early in his time in Oxford, Bernard also studied mathematics. Evidently he had a flair for this subject too; in 1669 the professor of astronomy – Christopher Wren – appointed Bernard to deputise for him in Oxford when he was appointed surveyor of the royal works. Four years later, Wren resigned the professorship altogether and Bernard became in name as well as fact the Savilian Professor of Astronomy.

He lectured on ancient mathematics and astronomy, was elected a fellow of the Royal Society and acted as tutor to a handful of minor noblemen. He published minor works on ancient weights and measures, on ancient languages and on etymology, and a collection of prayers. He catalogued the manuscripts held by English and Irish libraries. He was something of a manuscript-hunter, travelling to Leiden on a couple of occasions to acquire rare items at auction. True to the intentions of Henry Savile, Bernard also edited ancient texts. He was involved with drawing up an ambitious proposal to edit the entirety of ancient mathematics, in twenty-one volumes. He actually began an edition of the ancient geometer Apollonius, but he never finished it. He also worked for years on an edition of the Jewish historian Josephus. But his real dream was a new edition of Euclid.

Looking at the surviving evidence for Bernard's work on Euclid it becomes quickly clear why he was a man who published little and tended not to finish things. Bernard planned to compare the existing Greek editions with manuscripts, as well as correcting the best of the existing Latin translations against the Greek. And as well as a fresh, accurate version of the text, he intended to include the widest possible range of annotations. He planned notes on variant readings of the Greek text, cross-references to show the logical dependence of each proposition on others, and explanatory comments: all to be drawn from the riches of earlier printed editions and commentaries – Greek, Arabic, Persian and Latin – together with new information from Bernard himself and his contemporaries.

To this end, Bernard assembled a collection of printed copies of the *Elements*, as well as several manuscripts, and books containing the handwritten annotations of other scholars. On different printed copies of the text he wrote out parts of his carnival of Euclidean annotation, and he enlisted the help of his colleague, the Savilian Professor of Geometry John Wallis, to write and transcribe some of it.

Bernard was also interested in the Arabic versions of the *Elements*. There were manuscripts in Oxford, and he was able to draw on a printed source, too: a 1594 printing of a thirteenth-century Euclidean compendium attributed (wrongly) to the famous mathematician Nasir al-Din al-Tusi. Embodying the total of four centuries of study of the Euclidean text in Arabic, it was a storehouse of geometrical information and interpretation of the text, and Bernard chose it as the main basis for his study of the Arabic evidence.

It was perhaps inevitable that a project of such ambition would run into difficulties. There seems to have been some trouble coordinating the different layers of editing and commenting work, and Bernard's collection of printed versions of the *Elements* came to bear a complex tangle of kinds and styles of annotation. Some of them directly contradicted one another. More than once a whole series of meticulous annotations had to be just as meticulously crossed out.

Having filled the margins of (at least) seven printed copies of the *Elements* with preliminary work, Bernard turned to the mind-bending labour of collating it all into a single unified text which might be capable of being printed. To that end, he obtained four more printed books, at what must have been considerable expense: copies of the 1533 Greek edition, the 1594 Arabic edition, a 1612 Latin edition and a 1620 partial edition in both Greek and Latin.

He had all the pages taken out of their bindings, and he sorted the resulting mass of loose sheets together and had them re-bound so that he was left with, more or less, a single copy of the *Elements* in three languages, with the three texts running, as nearly as possible, concurrently on adjacent pages.

And he set to work to annotate this monstrous polyglot Euclid, putting in everything he had amassed so far: textual emendations and variant readings from the manuscripts, cross-references, endless detailed changes to wording, capitalisation, punctuation and even the placing of headings within the text, and of course commentary. Quickly the new copies themselves became something of a mess, with second thoughts and crossings-out, and several sections of the text never received certain of the layers of annotation at all.

The story now becomes a demoralising one. Bernard's friends repeatedly tried to discourage him from persisting with a project that had taken leave if not of common sense then at least of the reality of the book trade. One wrote to him that there was neither publisher nor market for such an edition, since most mathematical scholars did not know Greek: 'Men in this degenrous age wil not learne Greeke, or buy mathematicks att so great an expense of time and study and mony too.' Even if he could persuade a printer to print the book, Bernard would be left to pay the bills when it failed to sell.

Nevertheless, Bernard several times had specimen sections of the text printed, including extensive sets of diagrams. But by doing so he merely reinforced the point that the costs of the edition were likely to be terribly high, and his friends became blunter still, urging him to 'lay aside all thoughts of printing your Euclide'.

In the end, no publisher was found. Always in poor health, Bernard died of malnutrition and consumption in January 1697, his great Euclidean edition still no more than a mass of more or less disordered notes in the margins of printed books. He was buried in the chapel of St John's College, and the Bodleian Library paid his widow £340 for a selection of his books and manuscripts. Three hundred years later they remain in the library, the manuscripts unpublished and unpublishable as ever.

⬖

The cautionary tale has a coda, however. Six years before his death, Bernard was succeeded as professor of astronomy by a man twenty years his junior. David Gregory, educated in Scotland and a former professor of mathematics at Edinburgh, was a likeable and energetic teacher: not a brilliant mathematician, perhaps, but a sound communicator and author. He worked successfully with Edmond Halley on an edition of Apollonius: something Bernard had failed to complete. Later, perhaps inevitably, his attention turned to Euclid.

By 1698 senior figures within the university were writing to the well-known London publisher Jacob Tonson in support of a proposal to print the works of Euclid in Greek, promising that Gregory would 'take care of the Geometry & Reasoning'. The Greek text was to be overseen by one John Hudson, and there was further support from John Wallis, still in post as professor of geometry. Tonson had expertise in the printing of classical editions as well as being a founder of the notorious Kit-Kat Club.

No mention was made of Bernard, and the letters give the impression that this project was to be much scaled down compared with his ambitions. Tonson was, nevertheless, not impressed, and Euclid languished again. It was another five years before the Oxford University Press at last issued the complete works of Euclid in Greek and Latin. Gregory was credited as their editor, and in his preface he mentioned his debts to the manuscripts of Savile and the Bodleian Library, and the books of Edward Bernard. Some of Bernard's specific corrections to the text can be clearly seen in Gregory's edition, although the latter never acknowledged Bernard's long efforts to bring the text to publication.

Euclid's *Elements* had passed through a remarkable number of hands and been on a journey as long as that of any text, from the ancient world to the Renaissance. Around two millennia after its composition, and over two centuries after its first appearance in print – and after a multitude of more or less wayward printed versions – it had arrived at a point of rest, an edition with reasonable claims to be definitive: at least for the moment.

For all that, much of what Bernard had attempted to achieve was irretrievably lost in turning the edition into something that could practically be printed and sold. There was no Arabic in this edition; there were no lengthy explanations, references to other authors, variant readings or discussions of what was in the manuscripts. Gregory's Euclid was in most ways a pared-down edition, with little or no commentary and even some pruning of the text itself.

Athena in Oxford.

One of its few adornments was an illustration on the title page, showing Minerva with shield, helmet and spear seated against an Oxford background (she was sitting in the middle of what was and is Broad Street, holding up the traffic in fine style). In the background could be seen the old Ashmolean Museum and the Sheldonian Theatre where the book was printed: and, fittingly, the Bodleian quadrangle, housing the Savilian books and the Savilian study, where Edward Bernard had laboured for so long.

Interlude

Wait. Stop. All of these encounters with Euclid tell only one of the many possible stories about the *Elements*. They tell a story in which the *Elements* was a work of Greek literature, that travelled where Greek literature did, and was transmitted and translated just like Homer and Hesiod, Aeschylus and Sophocles. It is a story in which Euclid's was a text like other texts: beautiful, admittedly, but sharing a widespread tendency to become dirty, confused, corrupted under contact with human beings. It is a story, finally, in which Euclid is just an author like any other.

There was much more to it than that, though. People – a lot of them – have read the *Elements* not only for its own pure sake, but for what it could do for them: as a source of wisdom, or a spur to personal transformation. A different story would make the *Elements* a philosophy and Euclid a sage. It would show people reading the book because it put them in touch with eternal truths.

Back to Greece, then, to tell a different story this time.

II
Sage

Plato

The philosopher and the slave

Moonlight by the Mediterranean, in perhaps 415 BC. On the road between Athens and the neighbouring city of Megara, passers-by see a mysterious figure in women's clothing: a long tunic, a parti-coloured mantle and a veiled head. Even moonlight might reveal that it is no woman; those in the know recognise Euclid. His city is at war with Athens, barred to Megaran citizens. He takes the risk nonetheless, so eager is he to hear Socrates teaching and debating, and creeps into the city in disguise, night after night.

The veiled figure is another Euclidean legend, albeit one of the most attractive. Perhaps the most romantic of the tales about him, this one became associated with him in the Middle Ages. It rests on a case of mistaken identity, confusing the geometer with a philosopher who happened to share his name. Sadly, Euclid the geometer lived too late – much too late – to have known Socrates.

At the level of historical fact, the great Athenian philosopher

Plato – Socrates' most famous follower and spokesman – was indubitably interested in geometry. In perhaps the 380s or 370s BC, in his dialogue titled *Meno*, he presented a celebrated scene in which Socrates talked a compliant slave through a passage of geometrical reasoning.

'Tell me, boy, do you know that a square figure is like this?'
'I do.'
'Now, a square figure has these lines, four in number, all equal?'
'Certainly.'
'And these, drawn through the middle, are equal too, are they not?'
'Yes.'
'And a figure of this sort may be larger or smaller?'
'To be sure.'
'Now if this side were two feet and that also two, how many feet would the whole be?'
. . .
'Four, Socrates.'
'And might there not be another figure twice the area of this, but of the same sort, with all its sides equal like this one?'
'Yes.'
'Then how many feet will it be?'
'Eight.'
'Come now, try and tell me how long will each side of that figure be. This one is two feet long: what will be the side of the other, which is double in size?'

They draw a diagram, and Socrates asks a series of questions. Do you see . . . ? What do you think . . . ? Can you find . . . ? So, you say . . . ? The diagram shows a square, with a line falling across it diagonally. Under the philosopher's prompting, the slave shows Socrates how to draw two squares, one with twice the area of the other. The trick is to use the diagonal of the first square as the side of the new one.

The point, in the context of the dialogue, was to do with where knowledge came from. Socrates told the slave nothing – he merely asked questions – so the knowledge the slave manifested, he said, must have been recalled from a previous life. And 'if the truth about reality is always in our soul, the soul must be immortal'. The passage, deeply fictional though it is (no one really taught geometry to slaves), also happens to be the best, and virtually the only, evidence for the way people talked about geometry in ancient Greece. How geometry actually worked as live performance, as the drawing of diagrams and the making of deductions about them, a process of working through a proof with drawing stick in hand, trying to convince or explain or discover. Plato made it feel very much like Socrates' way of doing philosophy: a conversation in quest of the truth.

In another dialogue a decade or two later, Plato provided another equally celebrated geometrical passage. He used the regular solids – tetrahedron, cube, octahedron, dodecahedron and icosahedron – as the basic building blocks of the universe. His claim was that they were the shapes of the different elements – earth, air, fire and water – and that the properties of the regular solids could help to explain the properties and interactions of the elements.

Just what these passages, and others like them, mean, and how they should be read, has been much debated for nearly 2,500 years. It is clear, though, that Plato thought geometry important and that he, like other Athenian philosophers such as Aristotle, was willing to use it as a source of ideas and examples in his philosophy.

Not only that, but Plato's circle in Athens certainly contained some skilled geometers: indeed, they were presumably his sources of information for the geometry he used in his writings. Eudoxus of Cnidus, about a generation younger than Plato, worked on the theory of ratios. Theaetetus, a generation older than Plato, had one

Plato and his students.

of the dialogues named after him. He may well have been the first to discover geometrical constructions for certain of the regular solids – how exactly to build the solids out of circles, lines and triangles – and perhaps the proof that there are just five of them. He may well also have studied the various ways two lengths can bear a ratio to one another that cannot be expressed in whole numbers: like, for instance, the side of a square and its diagonal. All of this was material that, later, Euclid would collect into his *Elements*.

That said, it would almost certainly be wrong to imagine Plato himself doing mathematical research or taking any very close interest in it. Indeed, both Plato and Socrates had mixed feelings, on occasion, about theoretical mathematics. According to one contemporary, Xenophon, Socrates

> was against carrying the study of geometry so far as to include the more complicated figures, on the ground that he could not see the use of them . . . he said that they were enough to occupy a lifetime, to the complete exclusion of many other useful studies.

And Plato had reservations about the way geometers usually studied their subject. As he put it in his dialogue the *Republic*:

> Their language is most ludicrous . . . all their talk is of squaring and applying and adding and the like, whereas in fact the real object of the entire study is pure knowledge.

In other words, geometry was useful, but only if it was modified so as to point the student away from diagrams and their manipulation and towards the purely intellectual.

At the level of legend, however, Plato's interest in geometry tended to become somewhat exaggerated, with Plato as one of the original creators of the geometrical tradition: a sort of director of mathematical studies at his philosophical school, the Academy, in Athens, setting problems for mathematicians who eagerly studied them. The outcomes of this belief included the much later legend that the door to the Academy bore an inscription 'Let no one ignorant of geometry enter.' The first testimony to the inscription dates from the mid-fourth century AD, more than 700 years after the event, which places it firmly in the category of legend. In later Arabic versions the inscription became still more explicit: 'Let no one come to our schools, who has not first learned the *Elements* of Euclid.'

Again, once the mathematics of *Meno* and the construction of the five regular solids, and more, had become part of Euclid's *Elements*, the temptation to associate Euclid himself with Plato proved irresistible. The five regular solids were styled the 'Platonic' solids, their discovery ascribed (sometimes) to Pythagoras, and Euclid's placement of them at the end of the *Elements* was called a sort of homage to Plato's philosophy. Euclid now supposedly 'belonged to the persuasion of Plato and was at home in this philosophy'. And eventually the most picturesque of the legends about him was born.

The Athenians had provided in one of their decrees that any citizen of Megara who should be found to have set foot in Athens should for that suffer death; so great was the hatred of the neighbouring men of Megara with which the Athenians were inflamed. Then Euclid, who was from that very town of Megara and before the passage of that decree had been accustomed both to come to Athens and to listen to Socrates, after the enactment of that measure, at nightfall, as darkness was coming on, clad in a woman's long tunic, wrapped in a particoloured mantle, and with veiled head, used to walk from his home in Megara to Athens, to visit Socrates, in order that he might at least for some part of the night share in the master's teaching and discourse. And just before dawn he went back again, a distance of somewhat over twenty miles, disguised in that same garb.

As originally told, the story referred clearly to Euclid of Megara, a philosopher and student of Socrates born in the 440s or 430s: he lived in a different city and was of a different generation from the geometer Euclid of Alexandria. But in an accident that seems not quite an accident, it came to be widely accepted that the two men were one, that Euclid the geometer was also a philosopher, a member of Plato's circle, and had once disguised himself as a woman to enter Athens by night. Later medieval manuscripts and printed editions down to the late sixteenth century repeated the identification; a typical example – among hundreds – was the first Italian translation of the *Elements* which in 1543 credited its composition to 'Euclide Megarense Philosopho': Euclid of Megara, philosopher.

If Euclid was a Platonist, he did not say so; if he was keen on the use of geometry in education or philosophy, there is no evidence of it. He may never even have been to Athens. But the *Elements'* silence on every subject except geometry itself left its readers free to imagine Euclid's philosophical commitments, his historical situation and his biographical details as they pleased. And so the geometrisation of Plato, and philosophisation of Euclid, had consequences for how the *Elements* was read. The belief that Euclid had been a Platonic philosopher sanctioned later students to read his book as an expression of Platonic beliefs about a realm of perfect unchanging forms: a realm crammed with perfect geometrical shapes that no human could draw but any human – with the right training – could conceive of. To read the *Elements* in this mode, as an expression of a philosophy and a help to philosophical study, would become a 2,000-year tradition.

Proclus Diadochus

Minerva in Athens

A leafy estate near Athens, in AD 450 or so. You can see across to the Acropolis as you walk in the garden.

Proclus lives here; head of the Platonic school. A handsome, charismatic man, he lives a life of almost incredible discipline and intellectual labour. He rises at dawn to pray to the sun; sometimes he teaches five or more classes in a day, as well as writing typically 700 lines of prose. He visits other philosophers, and in the evening he lectures. His learning is encyclopaedic, his literary output vast.

Sunset, as well as noon, sees him bowing once more to the sun, and his reputation is for sleepless worship of the gods at night. He fasts monthly, abstains from meat, lives in constant communication with the gods in prayer and dreams. He knows the Chaldean rituals and the Egyptian; he celebrates the new moon with special solemnity. His prayers have been known to heal the sick and to bring about changes in the weather. He writes hymns.

> Great goddess, hear! and on my dark'ned mind
> Pour thy pure light in measure unconfin'd;—
> That sacred light, O all-protecting queen,

Which beams eternal from thy face serene:
My soul, while wand'ring on the earth, inspire
With thy own blessed and impulsive fire;
And from thy fables, mystic and divine,
Give all her powers with holy light to shine.

One of the surprises of the Euclidean story is that the first really substantial response to Euclid's *Elements* to have survived was written by someone ostensibly so unlikely to have been involved with the text at all. Proclus was the head of the revived Platonic Academy at Athens in the late Roman period, teaching philosophy to young men and women at the large private house that was attached to that role. It is not obvious that such a location would have seen any substantial engagement with mathematics; but in fact the philosophers of this circle saw their work as the correct interpretation of Plato and the harmonising of his writings with what was valid in a range of other sources, from Aristotle to the Chaldean oracles. They took very seriously the implications for mathematics in Plato's works; they took up, indeed, the belief that Euclid's *Elements* was an important tool of philosophy.

For Proclus there were three levels of reality. The One, apprehensible by the intellect; the physical world, about which human beings could learn, albeit imprecisely, by the senses; and, in between, mathematical things, which as it were looked both ways: they were eternal and unchanging and people could reason about them, but they shared with physical things characteristics like size and shape. This development of Plato's ideas (and there was a very great deal more to Proclus' metaphysics in this vein) meant that mathematics was crucially important in the life of the philosopher. It could sharpen the intellect in the practice of reasoning; it could also prepare and accustom the mind to the contemplation of the eternal and non-physical. More pragmatically, it could help in the study of physical things, as in physics or astronomy, where mathematics enabled precise reasoning about measurements and movements. Finally it could provide a model of valid reasoning that could be deployed in physics or theology: Proclus wrote an *Elements of Theology* and an *Elements of Physics* whose styles of reasoning were derived from that of geometry.

Proclus was distinctive in his special attention to geometry among the branches of mathematics. As he noted, unlike numbers, geometrical things shared qualities such as length, size and shape with physical things, strengthening the link and the analogy they bore with the physical world and making them more useful in sciences like physics as well as an easier 'bridge' for the mind in its ascent from the physical to the eternal. His teaching included regular lectures on geometrical subjects, and from that teaching arose his written commentary on the first book of the *Elements*. Proclus was not the first to write a commentary on part of Euclid's book, but his is the first to have survived in full and in Greek.

It has been called the most important philosophical document of its period, and it packs a great deal into a fairly small number of words. Two prefaces set out the importance of mathematics in general and geometry in particular for the student of philosophy. It is quite clear that Proclus as a teacher was used to dragging the unwilling through a 'geometry requirement' at the Academy, and to hearing complaints on the lines of 'I came here to study the great hierarchy of being, and you want me to draw *triangles*?' He disposed quickly of some obvious objections such as the apparent unimportance of certain geometrical results in themselves, or the unimportance of mathematics in practical life.

Proclus took a look at the history of geometry here, too. He had sources that have not survived to the present day, but he also had an agenda; his view was that the main stream of Greek thought was Platonism. Euclid, he decreed, had been a Platonist. Euclid, he claimed, stood at the end of a long line of geometers who wrote 'Elements' of geometry; he systematised their work, collected it and brought it into order, and he did so under specifically Platonic inspiration, placing the Platonic solids at the culmination of the book.

And, last, Proclus worked through the definitions and results of *Elements* book 1 themselves. He was writing less than a century after Theon, and he was a member of the same philosophical tradition as Theon's daughter Hypatia; he may represent a similar model of exposition to that of the Euclidean editor. It was detailed, even fussy, asking questions like: can this proof be broken by taking the lines and points in a particular configuration? Are there hidden assumptions? Does the argument lack certain steps? Can the proposition be split into separate cases, needing slightly different proofs? This, the main part of this book, was heavy work, and as a training for the powers of reasoning it adopted the approach of repetitive, unexciting practice. But the student who worked through it with full attention certainly had the opportunity to become dexterous at the type of geometrical thought it embodied, and to sharpen the geometrical imagination and the mind's quickness and flexibility.

To explain the geometric model for reasoning, Proclus also provided a schematic discussion of how a geometrical proposition was structured. As he conceived it, there were six parts to a proposition: enunciation, setting-out, definition of goal, construction, proof and conclusion. It was a neat scheme: far neater than the *Elements* itself, where the 'construction' section was often missing, as were the 'setting-out' and the 'definition of goal'; in other ancient mathematical texts Proclus' parts of a proposition could often scarcely be recognised. But, again, a system like this focussed students' minds on how proofs worked, on the geometrical method, which might be transferred to physics or theology or taken as a model of reasoning in general. Characteristically, Proclus showed how to break down Euclid's first proposition into its five parts, but suggested students should break down the other forty-seven propositions of book 1 themselves, as an exercise.

Writing a commentary of this kind naturally did something to Euclid's *Elements* itself. It did so in the most obvious sense that Proclus' commentary was long-lived, and later generations would insert fragments from it into the margins of the *Elements*, where they were transmitted along with the text and possibly sometimes were copied into the body of the text by mistake, as was the way of things.

It did something to the text in the subtler sense that it permanently changed how it would be read. Proclus inaugurated a long tradition of reading the *Elements* as a method, a logic, a way of reasoning. For him, its validity as a way of reasoning was a consequence of the place geometrical things held in his metaphysics, but in the long run people would find ways to drop the metaphysics and still retain his view of geometry as sharpening the mental powers and training the mind to focus on higher things. There is an irony to that, because, as many readers would find, the text did not really live up to Proclus' view of it. It was a real strain to make more than a handful of the propositions in *Elements* book 1 conform to his neat scheme of the parts of a proof, and later in the book the success rate was, if anything, lower. There were places where Euclid's proofs looked little like any tidy model. Furthermore, there were real logical gaps in the *Elements* and, as Proclus himself pointed out, hidden assumptions: such as that two straight lines cannot enclose an area. The struggle to fix these – to make the *Elements* what Proclus thought it was and should be – would become one of the most fruitful areas of research in Euclidean geometry over the centuries.

Hroswitha of Gandersheim

Wisdom and her daughters

WISDOM: O emperor, if you wish to know the ages of my children: Charity has completed a deficient, evenly even number of years; Hope a number also deficient, but evenly uneven; and Faith an abundant number, unevenly even.

HADRIAN: With such an answer you leave me not knowing what I asked.

WISDOM: That is not surprising, because under this kind of definition falls not one number but many.

HADRIAN: Explain more clearly, otherwise my mind cannot grasp it.

WISDOM: Charity has now completed two olympiads, Hope two lustres, and Faith three olympiads.

HADRIAN: And why are 8 (which is two olympiads) and 10 (which is two lustres), called deficient? Or why is 12 (which is three olympiads) called abundant?

WISDOM: Because every number is called deficient whose factors, added together, give a sum which is less than the number of which they are factors. Like 8: for the half of 8 is 4, its quarter is 2 and its eighth is 1; which, added together, make 7. Similarly with 10:

its half is 5, its fifth is 2, its tenth is 1; which, joined together, make 8. On the other hand, a number is called abundant when its factors added together exceed it. Like 12: its half is 6, its third 4, its fourth 3, its sixth 2, its twelfth 1, and the sum of these is 16. But I should not overlook the principle that between unruly extremes comes a harmonious mean; a number is called perfect which neither exceeds nor falls short of the sum of its factors. Such as 6, whose factors – that is, 3, 2 and 1 – make 6 itself. And for the same reason 28, 496 and 8128 are called perfect.

Books 7, 8 and 9 of the *Elements* dealt with numbers: odd and even, prime and composite, perfect and imperfect. In this as in most of the parts of mathematics on which he touched, Euclid provided an image of a tradition that was alive and active in his time, and both before and after it. And in this particular subject Euclid's is the only image to have survived from its period. The next extant full-length treatment of number theory is that of Nicomachus, from Gerasa in Syria, who lived around AD 100. (One of the very few authors known to history who worked on number theory during the 400-year gap was Hypsicles, as it happens: the one whose geometrical work became book 14 of the *Elements*.) Nicomachus' book was a proof-free handbook, a summary of well-known results, intended presumably for teaching beginners.

Roman statesman and translator Boethius gave a free-ish Latin version around AD 500, expanding the text and providing copious examples. In that form, as the *De institutione arithmetica*, the theory of numbers was widely known and studied in the Middle Ages. And so it turned up in the mouth of 'Wisdom' in the late tenth century. 'Sapientia' in the original Latin, she was a character in a play of the same name: a bloody tale of martyrdom and redemption. She and her three daughters were hauled before the Roman emperor on a charge of Christian proselytising (Hadrian was misrepresented, by the way: in reality he was tolerant of Christians). Failure followed failure for the would-be oppressor, who first tried without success to make the girls recant and worship the gods of Rome. Next he attempted to torture them, but divine intervention preserved them from feeling any pain: in the tradition of St Catherine of Alexandria, the instruments of torture variously failed, broke or turned upon those wielding them. The children having finally been beheaded and buried, Wisdom prayed by their grave and died in ecstasy, in a scene one critic has – perhaps somewhat optimistically – called a 'ray of Sophocles shining through a Christian mind'.

The number theory came near the beginning of the play, when Wisdom was goading the emperor, spinning out the answer to a simple question with a comically inappropriate explanation. Having made a fool of Hadrian with her defective, abundant and perfect numbers, she went on to explain the terms 'evenly even', 'unevenly even', and so on (which also went back to Euclid, via Nicomachus and Boethius). Finally:

HADRIAN: What a subtle and tangled question has arisen from the ages of these children!
WISDOM: In this we should praise the wisdom of the supreme creator, and the marvellous knowledge of the maker of the world. He not only ordered all things in number, weight and measure in the beginning, when he created the world out of nothing; he also in the following ages, and in the time of man, allowed the wondrous knowledge of the mathematical arts to be found.

Hadrian remained unimpressed, and perhaps a little humiliated by the display of this woman's knowledge.

Wisdom was written by Hroswitha of Gandersheim. She was possibly the most remarkable European woman of her generation, but scarcely more is known about her than can be gleaned from her plays and their prefaces and dedications. She was born probably in the 930s and spent most of her life as a canoness in the abbey at Gandersheim, a small town on the River Gander in Saxony. She was very likely of noble descent: admission to this convent was limited to the daughters of noble houses. She could well have entered the convent around the age of twelve.

Hroswitha as imagined in the Renaissance.

The Benedictine rule made for a communal life dominated by work and liturgy, although canonesses like Hroswitha were neither strictly cloistered nor forbidden to own property; they could receive guests, come and go subject to permission, and were even free to leave the convent permanently. The abbess was as powerful as a feudal baroness, with her own court, money, men at arms and a seat in the Imperial Diet. The convent at Gandersheim, founded late in the ninth century, had close ties with the Saxon royal house, whose court was for much of the tenth century a centre of learning and culture; and Hroswitha's education was evidently excellent, under – she reported – first one of the nuns, and later Gerberga, who was both abbess and niece of Otto the Great, Saxon king and Holy Roman Emperor. She learned Latin – her native tongue was Saxon – and read the classics; she learned philosophy, mathematics, astronomy and music. She may have learned Greek.

Her own writings included chronicles and sacred legends in verse, and six rhymed plays, of which *Wisdom* was the last. She was the first author to compose drama in Latin since antiquity; there would not be another until the mystery plays of the twelfth century. Her hybrid of saints' lives and theatre was prompted in part by the popularity of the pagan dramatists – notably Terence – as reading and teaching matter; Hroswitha set herself to provide a morally edifying alternative. Composing secretly at first and destroying some of her early efforts, she eventually produced a body of work that was admired by the learned critics she consulted at abbey and court, and that reflects an impressive light on the culture and library at Gandersheim. Her sources and allusions ranged across Roman and Christian authors: Horace, Ovid, Statius, Lucan, Terence and Virgil; Prudentius, Jerome, Alcuin, Boethius, Bede and many more, as well as the Latin Bible and the legends of the saints and martyrs.

She wrote about the history of her abbey and the Liudolf dynasty that had founded and supported it; she wrote about sin and virtue, including the earliest poetic treatment of a pact with the Devil. She wrote about saints and martyrs, and most particularly about women, whose potential for courage, grace and spiritual energy she depicted in vibrant colours. Her gift was for a striking presentation of the great moments in a story, and she had a taste for humour too; a much-quoted passage had a lecher, Dulcitius, embracing a variety of kitchen utensils under the delusion that they were girls. Frequently the plays showed verbal incongruity as wit: small talk drawn out, explanations that failed to explain, incongruous displays of erudition.

To display her own erudition was also important to Hroswitha, and as well as the numerical passage in *Wisdom*, others of her plays contained discussions of musical theory and of logic, testifying to her wide reading and thorough education. She loved to coin new words or to find new meanings for old ones. Her verse amounted to an adaptation of the classical metres – hexameters and pentameters – to the pronunciation and the preference for rhyme of her period. As she put it in one of her prefaces, God 'has given me a perspicacious mind, but one that lies fallow and idle when it is not cultivated'.

Hroswitha died probably around AD 1000, leaving an intellectual legacy remarkable by any measure, and a unique record of reading and learning by an early medieval woman: reading and learning that included among many other things the Euclidean number theory in Boethius' Latin version. It is not known whether her plays were performed in her lifetime, although they were certainly read well beyond the walls of the convent and quite probably at Otto's court. The literary prestige of her monastery declined after the end of the tenth century, but enough manuscripts of Hroswitha's works continued to be produced to suggest an ongoing interest in her through the later Middle Ages.

Around 1500 a German scholar and poet discovered manuscripts of Hroswitha's works in the Benedictine monastery of St Emmeram at Ratisbon, and published them in 1501. Since then she has attracted scholarly and popular attention: her plays were performed

– possibly for the first time, as far as the evidence goes – in the late nineteenth century, and from 1944 to 1999 the New York-based Hroswitha Club sponsored scholarly research about her and her works among its other bibliophilic activities. Performances continue to take place, and 1,000 years after her death her extraordinary mind has perhaps not yet reached its highest point of visibility.

Rabbi Levi ben Gershom
Euclid in Hebrew

The town of Orange in France, famed for its preserved Roman amphitheatre; the late 1330s.

In the Jewish quarter, a rabbi writes in Hebrew. 'It has seemed good to us to complete what needs to be in the book of the *Elements*, since this book is of the greatest profit for geometry, for which it provides the fundamental principles.'

Euclid's *Elements* did not begin with geometrical propositions. It began with a series of prefatory statements, including definitions and basic assumptions. After the definitions, saying what was meant by a point, a line, and so on, there were what the text called 'postulates' and 'common notions'. 'Common notions' were things like 'the whole is greater than the part': not really specific to geometry, and for most people so obvious that they hardly needed to be said. The 'postulates' were the assumptions that it was possible to draw such things as straight lines and circles at all: again, assumptions so obvious that many readers did not feel they even needed to be stated.

The postulates and common notions were the subject of a huge amount of editorial intervention over the centuries, as editors tried to arrive at a list of basic assumptions that was reasonably complete without containing too much of the very obvious. Their number varied from manuscript to manuscript, as did their classification, sometimes involving the separate category of 'axioms'. There is still real uncertainty as to which, if any, were put there by Euclid himself.

There were specific problems with one of the postulates, and they would dog the entire history of the *Elements*. Postulate 5 said that:

> if a straight line falling on two straight lines make the interior angles on the same side less than two right angles, the two straight lines, if produced indefinitely, meet on that side on which are the angles less than the two right angles.

Perhaps the kindest remark that was customarily made about it was that it was strikingly inelegant. Just to work out what it meant was an effort. It captured the intuition that if two lines were not parallel, they would meet somewhere: but it captured it in what seemed a wordy and hard-to-use way. It came to be called the 'parallel postulate', and almost every editor of the *Elements* had something to say about it.

Some tried to replace it with a simpler or more easily comprehensible equivalent; some tried to prove it from more basic assumptions. Such attempts began at least as early as the first century BC and continued through the work of geometers in Greek, Arabic, Latin and beyond. Proclus reported a couple of false proofs from the Greek tradition: the usual trouble was that the geometer turned out to have assumed, tacitly, something equivalent to the parallel postulate itself. Arabic attempts, from Euclidean translator Thabit ibn Qurra in the ninth century to commentator Nasir al-Din al-Tusi in the thirteenth, were also failures.

One of the more determined and interesting attempts to solve the problem of the parallel postulate in the Middle Ages appeared not in Greek, Arabic or Latin but in Hebrew, at the hands of Rabbi Levi ben Gershom of the town of Orange.

Jewish culture had flourished under Islam in the Iberian Peninsula, adopting Arabic as a vernacular and developing philosophical and theological traditions in contact with Arabic writings and writers: and through them with Greek sources such as Aristotle. A sense grew that Judaism could enrich and strengthen itself by the study of Aristotelian logic, Aristotelian physics, even Aristotelian metaphysics. For the celebrated thinker Moses ben Maimon (Maimonides), writing his equally celebrated *Guide for the Perplexed* in Arabic in the twelfth century (it was later translated into Hebrew), acquiring knowledge was a necessary part of the perfection of the soul.

But from the 1140s onwards the Jews were driven out of Andalusia in the wake of the Almohad invasion. They moved into northern Spain and southern France, where the Comte de Provence and the Prince d'Orange were sympathetic to immigrants generally; and the papal court, itself in exile in Avignon, was willing to grant status and patronage to the learned of any religion. By the fourteenth century, Provence had 15,000 Jews among a population of 2 million, working as moneylenders, physicians, craftsmen and merchants. Sizeable towns like Orange and Avignon had their own Jewish quarters and Jewish cemeteries.

Translation into Hebrew became important, as the Jewish communities' use of Arabic declined: scores of works by Greek and Arabic authors were rendered into Hebrew, in a moment of cultural engagement and assimilation unprecedented in Hebrew literature. A large body of learned writings became available in

Hebrew, and a scholar who read only that language now had the possibility of doing original work in a range of fields from logic and mathematics to astronomy and metaphysics.

Of course, inevitably, the *Elements* was translated too, among a selection of other mathematical works. No fewer than four Hebrew versions of the *Elements* were made from Arabic sources during the thirteenth century, as well as one or more from Latin and two from Persian. The *Elements* rapidly became not just the most translated but the most commented on and the most copied mathematical work in Hebrew. New commentaries were written in Hebrew – at least fifteen – and a number of Arabic commentaries were translated. As at other times and places, the *Elements* became one of the works the educated and cultivated wished to have on their shelves, one of the learned books that turned up in the libraries of intellectuals in all fields.

Levi ben Gershom (or Gerson) was Leo de Balneolis to the Latins, or Leo Hebraeus, or Gersonides; in the conventional Jewish acronym he is known as Ralbag (from **R**abbi **L**evi **b**en **G**ershom). He was born in 1288 and died in 1344. A member of the apparently learned and certainly important de Balneolis family, he most likely lived his whole life in Orange, in the city's Jewish community of perhaps fifty or a hundred families. He spoke Hebrew and Provençal; he wrote only in Hebrew. He may have known some Latin or Arabic, but the evidence is slight.

His father and grandfather seem to have been Talmudic scholars, but the details are uncertain. A brother was a doctor. It seems he married a cousin, but it is uncertain whether there were any children. It is not known, in fact, how ben Gershom made a living; scanty evidence connects him with medicine or banking (money-lending), or with the traditional function of a community rabbi.

It is certain, though, that he became the leading intellectual of his generation: a major philosopher and one of the pre-eminent medieval scientists. He took the Arabic–Hebrew philosophical tradition forward, striving to work out a physics and a metaphysics which would reconcile the core doctrines of Judaism with Aristotelian science and philosophy. Ben Gershom wrote commentaries on many of the Aristotelian works of twelfth-century Muslim philosopher Ibn Rushd, known in the West as Averroës; he also wrote a series of commentaries on the Hebrew scriptures. He produced original works and commentaries on logic, and finally his great work the *Wars of the Lord*, which has been called the most sophisticated work of theology in the history of Judaism.

He was a passionate believer in the power of human reason to discern the truth: to gain true knowledge of real things by means of the evidence of the senses, and to perfect the sciences over time. He undertook empirical science to an extent unparalleled in Hebrew scholarship; in particular, he was the most important observational astronomer of his generation, perhaps of his century. Never content to rely on data in books if he could improve on it himself, he was observing the heavens by 1320 and continued to do so on and off for the rest of his life. As well as using the camera obscura, which had existed in Europe for half a century or so, he invented a new instrument, known today as 'Jacob's staff' due to the misunderstanding of an allusion to the patriarch Jacob in one of ben Gershom's poems. The 'staff' consists of a long stick, with a short stick mounted at right angles and able to slide along it. With an eye at one end of the long stick, the observer points it at the sky and adjusts so that a star appears at each end of the short stick. Noting the position at which the short stick stands, and knowing its length, it is possible to work out the angle between the two stars in the sky. It may sound rudimentary, but it was probably the most useful astronomical instrument before the telescope (Copernicus used it), and it would have a long afterlife as a navigational instrument, in the sixteenth century and later. Based on his observations, ben Gershom revised the accepted system of the motions of the planets, massively increasing the distances to them and coming up with a much simplified model for the moon's motion.

His work made him a celebrity. There are passages in ben Gershom's books about how engaging in philosophy can raise one's status, and it undoubtedly raised his. There are passages about how to deal with princes, and he had certainly learned to do so. Frustratingly little is reported about his relationships within the Jewish community, but ben Gershom is known to have had dealings with prominent Christians, including at the papal court. Aged thirty-three, he computed an almanac for finding the date of the Christian Easter, at the request of 'many great and noble Christians'; there were both Provençal and Hebrew versions. Later, in the 1330s, astrological predictions were commissioned by Pope Benedict XII, and in the final few years of his life ben Gershom reworked his *Astronomy* and dedicated to Clement VI a Latin translation of parts of it. It seems he was to some degree a client of the papal court.

Plato and Proclus, roughly speaking, had believed one should study mathematical things because they lived in the realm of eternal forms and could lead one away from the mundane and towards yet higher things. Ben Gershom did not believe in a world of forms, but thought that mathematical things, like other truths, were abstractions from sense data. Nevertheless, he subscribed to the value of mathematics in a philosophical education. How, indeed, to make the best use of empirical observation, so as to improve one's own intellect and the state of human knowledge? He thought a person should study mathematics because it would help to form abstractions correctly, to spot and work with the true patterns of the world, particularly in a science like astronomy. Thus, ben Gershom's ideal education found a place for mathematics at its very beginning, as a preparation for what was in this view the higher study of physics and astronomy (which in turn was a preparation for the study of metaphysics). With this in mind, he wrote a book on arithmetic and algebra, in a somewhat Euclidean style, and one on trigonometry.

And so Levi ben Gershom turned to Euclid, to which he brought the eye of both the philosopher and the astronomer. He regarded the arithmetical books of the *Elements* – the very parts Hroswitha had made the daughters of Wisdom quote – as prerequisite reading for his own arithmetical work, in which he referred to them. He regarded the geometrical parts of the *Elements* as necessary for the proper study of astronomy.

But certain things troubled him, all the same. Like others, he found that for the purposes for which he was reading geometry, Euclid's definitions needed to be improved, clarified and expanded, and he set himself to write a commentary on the book, providing that expansion and clarification. For Euclid's 'a point is that which has no part' he noted that other commentators had already noted that a point needs to be distinguished from the arithmetical unity by saying it has a position. To Euclid's definition of the diameter of a circle he added a proof that the diameter thus defined really does cut the circle into two equal parts. When Euclid postulated that a given straight line can be indefinitely extended, ben Gershom again found himself troubled. Aristotle had pointed out that unless the universe was infinite, no line could really be indefinitely extended; and ben Gershom found it necessary to distinguish between the physical and the mathematical aspects of a line. And so on, through some of the other definitions in books 1 to 5 of the *Elements*.

Turning to the parallel postulate, that baffling chunk of prose about the behaviour of intersecting lines, ben Gershom not surprisingly found it wanting, as many had before him. He was aware of the work of earlier commentators on the subject, and he wrote about it at length himself. His idea was to replace the offending postulate with two others, which he reckoned more self-evident, and then use them to prove the parallel postulate itself. His replacement postulates were these, hardly models of lucidity and comprehensibility:

that a straight line can be extended to a make it greater than any given straight line; that if two given lines form an acute and a right angle, respectively, with a third line that cuts across them, then they grow closer to one another on the side of the acute angle and grow farther apart in the opposite direction.

His proof then involved a chain of nine subsidiary results before arriving at the required parallel postulate, now redesignated as a theorem. He thus displayed his own dexterity at Euclidean geometry – which was considerable – but as far as the foundations of geometry were concerned he had unfortunately achieved little more than to replace one doubtful thing with another.

Ben Gershom himself was not wholly satisfied with his attempt to fix the *Elements*, and he came to feel, in fact, that geometry needed a whole new set of foundations; an improved presentation which would avoid some of the gaps and difficulties he found in Euclid's book. So he set out to write his own treatise on geometry, putting the subject on a stronger foundation. In it, he brought together the ideas from his commentary on the *Elements*, improving the same definitions and proving the same assumptions. The important aim of proving the parallel postulate was the same.

Ben Gershom's new geometrical treatise did not catch on: today just one manuscript of it survives, containing the first twenty-four definitions and hypotheses. The *Elements*, instead, continued its long reign over both mathematics and – as ben Gershom had himself endorsed – philosophical education. He was not the last to trouble over the adequacy of Euclid's definitions, nor to worry about the status of the parallel postulate and whether it might be proved from more self-evident premises. Nor was he the last philosopher to agree that geometrical knowledge was structured, broadly, as Euclid had structured it, but to doubt whether the detailed contents of the *Elements* were quite right.

Christoph Clavius

The Jesuit *Elements*

A small book was printed in 1574; a little smaller than a modern paperback, but with hard covers and about 700 pages. Today, its binding is rather torn; someone has scrawled 'EUCLID' on the spine, together with a shelfmark ('L 135'). The flyleaves bear the signatures of at least three previous owners; they are partly crossed out, making them hard to read. The book was part of an institutional library at one time; the words '. . . of St Francis' appear in one place.

The book is the first edition of Euclid's *Elements* in the version by Christoph Clavius of the Society of Jesus: or rather, the first of the two volumes of that first edition. The individual volumes would fit in a large pocket or a small bag. The type is small, and there is a great deal of text here; the introduction makes the almost unbelievable claim that the edition contains 1,234 Euclidean demonstrations: more than three times as many as in most versions. The book incorporates huge quantities of notes and additions drawn from all kinds of sources: alternative proofs, alternative propositions, alternative axioms, proofs of the axioms, refutations of false proofs . . .

Christoph Clavius was born in the small German town of Bamberg, on the feast of the Annunciation in 1538. Two years later the Jesuit order – the Society of Jesus, founded by Ignatius of Loyola – received papal approval, and young Christoph was most likely won over by one of its visiting preachers. He entered the Society shortly after his seventeenth birthday, and went to study at the University of Coimbra in Portugal. He would never return to his home town.

Coimbra had at least one distinguished mathematics teacher, but it is not clear whether Clavius learned from him or from books. By whatever route, though, he became rapidly expert in the subject. He returned to Rome for advanced theological study in 1561, and within two years he was appointed the mathematics professor at the Jesuits' college there. He remained at the college – the Collegio Romano – for the rest of his long life, barring a few periods of travel.

Ordained in 1564 and fully professed in the Society in 1575, Clavius became a respected authority on mathematics, a prolific author and a powerful teacher. Contemporary testimony makes him

> a man untiring in his studies and . . . of a constitution so robust that he can endure comfortably the long evenings and efforts of scholarship. In stature he is well proportioned and strong. He has an agreeable face with a masculine blush, and his hair is mixed in black and white. He speaks Italian very well, speaks Latin elegantly, and understands Greek. But as important as all these things, his disposition is such that he is pleasant with all those who converse with him.

One of his pleasantries is recorded. Asked by Pope Gregory XIII whether he had good living quarters, 'comfortable and suited to

Christoph Clavius.

his studies', Clavius replied 'Good? The best! . . . All I have to do
is move my bed from one room to another when it rains at night
so that the water doesn't fall on my head.'

He carried out astronomical observations; he had the rare good
fortune to witness a total solar eclipse at Coimbra and another at
Rome, as well as a number of novae. He was one of the first
generation to look through a telescope, and he saw what Galileo

saw: the phases of Venus and the markings on the moon. He published astronomical textbooks and books about scientific and mathematical instruments, and was involved with the new 'Gregorian' calendar, defending it against its critics; it came into use from 1582 onwards, and is still in use today.

One might reasonably wonder what the Society of Jesus wanted with a mathematical expert of this calibre. The Society was founded with an ideal of itinerant ministry, but within the twenty years that saw the vocation and training of Christoph Clavius it moved more and more towards educating children and adolescents. When he joined, there were about 1,000 Jesuits running thirty-odd schools; by the end of the sixteenth century there were over 8,000, and nearly 250 schools. The shaping of their curriculum became an important question, and in the years leading up to the issue of a definitive programme of studies in 1599 Clavius was involved with agitating for mathematics to be given an honoured place on it. An informal mathematical academy within the Collegio Romano existed to train technical specialists (architects, surveyors, administrators), to give future missionaries the scientific expertise they might need in remote locations, and to train teachers for the Jesuit schools. The scarcity of mathematics teachers was indeed a recurring problem for Jesuit schools throughout the first century of the order's history.

So far went the pragmatic arguments; but the status of mathematics also continued to be a more fundamental philosophical issue. The Platonist view of Proclus and others had been widely read and taken up into strand after strand of medieval Latin thought. St Augustine, in particular, transmitted to Christian philosophy the general notion of geometry as a specimen of the eternal and unchanging. The Euclidean translators Boethius and Adelard were both interested in and familiar with Platonic ideas. Aquinas praised mathematics; Albertus Magnus wrote a Euclidean commentary.

On the other hand, the Latin world's rediscovery of Aristotle in the twelfth century had pulled in another direction. For Aristotle, there was no realm of eternal forms; mathematical ideas, like other ideas, were abstractions from the evidence of the senses (as they were for ben Gershom), and they had no other reality. Thus mathematics had no particularly exalted status. Furthermore, Aristotle's much-studied body of works on logical proof privileged verbal reasoning in the form of syllogisms. All men are mortal, all Greeks are men, therefore all Greeks are mortal: and so on through thirteen other combinations of premises and conclusion. If this was the road to certain knowledge, the alternative model of Euclidean deduction was an irrelevance and a distraction. Mathematics showed a tendency to lose its status in the medieval universities as a result.

By Clavius' time, opinions were swinging back in Euclid's favour. The works of Plato had been rediscovered and translated from the fifteenth century on; the works of Proclus were published in the sixteenth. Some mathematicians felt emboldened to push back against the Aristotelian line about both what mathematics was and what it was good for, arguing for a higher status for the discipline both socially (for its practitioners) and in the hierarchy of forms of knowledge (for mathematics itself); they deployed largely the same arguments that Proclus had used. It was also the case that the practical usefulness of mathematics was becoming more visible, more regularly acknowledged by intellectuals.

Mathematicians, in short, had scented the real possibility – after several centuries in the shadows – of boosting the status of their discipline and incidentally their own status, by appealing to writings in the Platonic tradition that said mathematics was basic to knowledge and reality. Copernicus, Galileo and Kepler, to name just three, were all steeped in Proclus' commentary on Euclid. Ratdolt had shown by his innovative printed diagrams how geometrical reasoning – even in print – could be truly exact, not merely approximate, and an influential Latin translator of Euclid from the Greek around 1500, Bartolomeo Zamberti, added a preface that was all about directing the mind to the immaterial through the use of geometry.

All of this gave a new urgency to certain kinds of work on the Euclidean text that had been done, on and off, since antiquity. A writer who wished to claim Euclid's *Elements* as a superior equivalent to Aristotle's logic needed to be sure its logical content really was correct. Editors turned with new attention to tasks such as checking proofs for faults or gaps, editing them into a transparently consistent structure and making sure they covered every case they were supposed to cover and made no awkward unstated assumptions.

This last became a matter of particular anxiety, and through the sixteenth century edition after edition gave a fresh or improved set of postulates and common notions, to try to make the *Elements* as logically irreproachable as could be. Over the course of the sixteenth and seventeenth centuries more than 200 different postulates or axioms were tried in various combinations; some editions had nearly fifty. Starting from Euclid's 'the whole is greater than the part', for instance, editors in the sixteenth century tried to define terms ('A whole is what has parts'; 'Every whole is divided into parts') and to extend and refine the axiom ('The container is greater than the contained'; 'The measure is not greater than the measured thing'). Noticing that there was no axiom requiring crossing lines actually to have a point of intersection, some supplied one ('If a straight or curved line is drawn from a point which is within a figure to another point in the same plane which is outside the figure, it will intersect the sides or boundary of the figure'). Others again attempted gap-filling refinements of Euclid's 'If equals be added to equals, the wholes are equal', such as 'If unequal things are added to unequal things, the greater to the greater and the lesser to the lesser, the wholes will be unequal, and in fact the former greater than the latter.'

It was in this context that Clavius issued his edition of Euclid's *Elements*. It was nearly a century since Ratdolt had first put the text into print, and Clavius was the heir to a long, rich tradition of work on the text. He sought to synthesise what was best in the many editions of his predecessors, presenting material from the ancient, medieval and sixteenth-century versions of the book. And he added numerous corollaries, lemmas and comments of his own, and even some new propositions. To Euclid's five postulates and five common notions Clavius had four postulates and twenty axioms, for instance, and he gave no fewer than 585 demonstrations additional to those in the Euclidean text proper. Thus his book became a packed, teeming, tangled space, dense with notes, reminders and cross-references, and with the boundary between text and commentary constantly in danger of breaking down. It came close to being a new book altogether, and it richly illustrated that Euclidean geometry was still a living, growing thing, and hinted that the able student might reasonably expect to contribute to it too.

His preface hammered home a message about the importance of mathematics. Clavius cited Plato on the importance of mathematics to philosophy; he drew heavily on Proclus, and modelled the preface to his *Elements* on the preface provided by Proclus' German editor. He cited the whole tradition from Augustine and Jerome downwards for the use of mathematics in Christian theology: 'nobody can accede to metaphysics if not by way of mathematics'.

But he also emphasised the potential of mathematics for real practical uses. Clavius the astronomer and calendar reformer was in a strong position to make these arguments, and there were subtle hints throughout his version of the *Elements* that he wished to facilitate such applications as far as he could. His diagrams emphasised the physicality of the things they depicted. He included diagrams for making fold-together models of the Platonic solids, so that the interested reader could build paper versions to handle. So convincing was his defence and celebration of mathematics that Clavius would reuse this preface wholesale four decades later, as the preface to his own collected works.

So, Clavius united two approaches to mathematics: the theoretical and the practical. One that saw geometry as a route to universal truths; one that saw it as a practical art, a tool for real-world ends. Clavius stood at the intersection of two approaches to mathematics in culture and in education, and by doing so he created something of a classic, a version of the *Elements* that united the two strands – intellectual and practical – that would make up the Scientific Revolution of the seventeenth century. His book was revised and republished five times in his lifetime, and printed again in the mid-seventeenth century. A string of editions by notable Jesuit mathematicians followed his lead, taking the tradition he had started down into the later seventeenth and early eighteenth centuries. His admirers called him 'the Euclid of his times'.

His book crossed international and confessional boundaries: copies of Clavius' Euclid were donated to the (Protestant) colleges of the University of Oxford in the seventeenth century, for the use of bright graduate students there. The edition was a reference point for any discussion of the nature of Euclidean geometry for perhaps 200 years; it has been called – plausibly – the most important edition of Euclid ever published.

In the context of Jesuit education, too, Clavius' version of the *Elements* gained immense importance, and he largely won his battle for the status of mathematics and geometry within the order. The informal mathematical academy at the Collegio Romano gained official recognition in about 1593, while Clavius himself had enjoyed the title and dignity of a professor for twenty-five years. The programme of study he drew up for the more able mathematicians there included the whole of Euclid's *Elements*, interspersed with other studies in arithmetic, trigonometry, and the applications of mathematics such as astronomy and musical theory (Clavius himself wrote songs and motets). In this curriculum, the *Elements* was, naturally, the first and most important book to be studied. The official Jesuit programme of studies issued in 1599 prescribed the study of Euclid in all Jesuit schools, as well as the teaching of mathematics to advanced students of both physics and philosophy. Although more ambitious proposals with more binding requirements as to the amount of mathematics to be studied were quashed, this was a decided victory for Clavius' view of the importance of mathematics in Jesuit education and the place of Euclid within the mathematics course. By the early seventeenth century the mathematical disciplines had attained pretty well the status Clavius had wished for them. Near the end of his life, indeed, a special school to train Jesuit mathematicians was set up in Antwerp; one of the early masters was Gregory St Vincent, who had been a student of Clavius in Rome.

Thus the nearly a quarter of a million children and adolescents in Jesuit care at any given moment would now be exposed to Euclid's *Elements* in Clavius' version; and as well as Clavius' students at Rome and the long line of impressive Jesuit mathematicians, the order would also go on to educate such non-Jesuits interested in mathematics as Descartes, Laplace, Diderot and Voltaire. Clavius' love for Euclid, and his distinctive vision of the *Elements*, would shape mathematical culture for two centuries.

By the end of his life Clavius himself was something of a celebrity, and visitors to Rome sought him out. One was Galileo, in whose telescopic observations Clavius had taken an interest. He reckoned the venerable Jesuit 'worthy of immortal fame', and it has long been suspected that Galileo both used Clavius' *Elements* in his own teaching and took from the older man some inspiration for the role of mathematics in natural philosophy.

Some of the contemporary praise of Clavius is mere hyperbole, or misses the point. But he might have been pleased by one detail. Someone at Cambridge early in the seventeenth century wrote on the title page of a copy of Clavius' *Elements*: 'here are the wonders of God and the mysteries of the world'.

Xu Guangqi

Euclid in China

Beijing, in the thirty-third year of the Wanli reign; AD 1604.

Xu Guangqi, in the capital for his Civil Service exam, visits his acquaintance Li Ma To. A wise man from the far West, Li has travelled a year to come to China, from his home at the far end of the world. He has a reputation as a man who never lies, an elegant speaker and skilled disputator and the possessor of a prodigious memory. He has published books in elegant literary Chinese on friendship, on the art of memory and on the new religious cult he brings with him. His Chinese home is stored with mysterious artefacts: instruments, maps, treasures. And books.

Xu Guangqi was the son and hope of an urban family in Shanghai. Flood, drought, famine and the steady erosion of his father's inheritance meant that a Civil Service career – entry to the elite class through erudition and culture – was his best option. He passed the local exam, the *xiucai*, in 1581, failed the regional *juren* exams three times over the next decade and finally achieved his degree in 1597, having worked as a teacher and tutor mean-while. It took him seven more years to pass the final *jinshi* degree in Beijing.

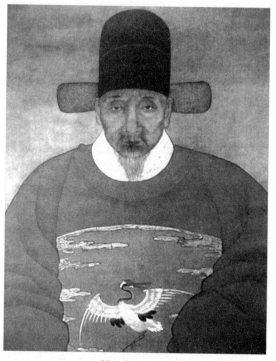

Xu Guangqi.

Like many of his class and time, Xu was frustrated by what he felt was the sterile study of literary and philosophical classics, the basis of the mandarin exams. The crises of his early life and the threat of the Manchu on the northern border added to his impa-tience with mere speculation. A Confucian, he wished to see learning yield practical results in the service of his country. He developed interests including water engineering, and in 1603 he presented the magistrate of Shanghai county with a text outlining methods for surveying rivers.

His life changed in the most unforeseeable way during the years around 1600. In the 1590s he encountered a man named Cattaneo in Shaozhou: a traveller from the West promoting a religious 'way' distinct from the Confucian, Buddhist and Taoist traditions known to Xu. In 1600, en route to Beijing for his second unsuccessful try at the *jinshi* exam, he met Li Ma To in the southern capital Nanjing: another of the same sect, who made an enormous impression on him. Xu would write that 'this was the only gentleman in the world who understands the relationships between all things'.

Years of reflection in Shanghai brought Xu back to Nanjing in 1603, but Li Ma To had left. His colleague João da Rocha received Xu, taught him and gave him reading matter including Li's manuscript treatise on the 'Lord of Heaven'. Xu found himself sufficiently convinced by the 'way of the Lord of Heaven' to accept the rite its followers called baptism, again at Nanjing. He took the name Paul.

In the *jinshi* examination the following year Xu failed at the first scrutiny, but was later passed when the examiners took a second look at his paper. He saw this success as a mark of divine favour, and renewed his contact with Li Ma To, who was now living in Beijing on an imperial stipend.

Li Ma To was a man literally from the far side of the world. His native town was Macerata in the Papal States, and his name in the West was Matteo Ricci. Ten years older than Xu, he had trained in Rome as a lawyer and there joined the Society of Jesus. Study at the Collegio Romano, where his masters included Christoph Clavius, was followed by his volunteering for the Far Eastern missions. A predecessor, Francis Xavier, back in 1552, had failed even to enter the Chinese mainland, but later that decade Portuguese Jesuits had visited Guangzhou and it seemed that sporadic contact was now likely to flower into a permanent missionary residence on the mainland. Ricci and seven companions left Rome in 1577.

They travelled via Portugal – where there was further training at the University of Coimbra, including mathematics under a student of Clavius – and Goa, which meant a year at sea rounding the southern coast of Africa. It was one of the most ambitious journeys a person could undertake in the sixteenth century, and for Ricci it proved to be a one-way trip. Twenty-four years old, he had not seen his parents in nine years and would never see them again; after leaving Portugal he never returned to Europe.

After Goa – where Ricci finished his theological studies – and now ordained, he moved on to Macao, the next in the chain of Portuguese colonies. There he began to learn Chinese with Michele Ruggieri, another Italian and the founder of the Jesuit mission in China. And after Macao, Zhaoqing. This was in 1583.

There was move after move within China, as the Jesuits pushed forward towards their ultimate goal of the imperial city and negotiated the ebb and flow of official favour and protection: or of suspicion and, sometimes, expulsion. Travel seems to have agreed with Ricci, and much of what he knew of China he learned on his journeys through the country's landscape: its rivers, lakes and canals. He noted distances and locations and the details of life along his routes.

Ricci and his companions were based in Shaozhou for twelve years; they visited Nanking. They spent three years in Nanzhang and, after a failed first attempt, moved permanently to Beijing in 1601. Once again, there would be no turning back; Ricci would never again leave the imperial city.

In Beijing the mission was initially accommodated (detained might be a better way of putting it) in the residence for foreign envoys. But later the Jesuits were allowed to rent and eventually to purchase a house, supported materially by an imperial stipend and socially by the high-ranking officials who took an interest in them.

Ricci was enormously impressed by the architecture of the

Chinese capitals Nanjing and Beijing: the walls, the troops and the fortifications. He admired the dignity and power of the mandarins and the grandeur of the inaccessible imperial court. He took steps to impress on his own account, too. It had always been the intention of the mission to blend in, to adopt local customs and styles of dress as far as possible. Initially, acting on what turned out to be poor advice, they dressed as Chinese monks, with shaven heads and chins and quasi-Buddhist robes. But being perceived as a new kind of Buddhist, they found, did little to help their cause, and by the 1590s Ricci and his companions had grown out their hair and dressed themselves in full Confucian regalia. His description gives an unmistakable sense that he enjoyed his new image, with robes

> of purple silk, and the hem of the robe and the collar and the edges . . . bordered with a band of blue silk a little less than a palm wide; the same decoration is on the edges of the sleeves which hang open, rather in the style common in Venice. There is a wide sash of the purple silk trimmed in blue which is fastened round the same robe and lets the robe hang comfortably open.

As an exercise in image manipulation, it worked, and the Jesuits now looked to their hosts like worthy peers of the erudite elite. Peers, for instance, of Xu Guangqi.

What did Xu and his compatriots see in Matteo Ricci who was Li Ma To? Men from the far West, who had travelled a year – perhaps many years – to get to China (true). Whose companions had died along the way (legend, but widely reported). Whose own land was a peaceful utopia full of marvels (also legend). Men who spoke Chinese well enough to be understood, but whose own languages were strange.

The Jesuits had arrived laden with a fantastic collection of arte-facts: clocks, prisms, astrolabes, quadrants and globes. They had books printed using movable type, engravings and oil paintings in perspective. They had mirrors, linen cloths and a clavichord. They had a map of the world (this was particularly admired). Some of these items were intended as gifts for the emperor; some were put on display in the mission residence. The 'cabinet of curiosities' was a well-established institution in the Europe Ricci had left behind, and the Jesuits managed to recreate something like it in their house in Beijing.

All of these 'wonders' were conversation-starters, as of course were the Jesuits themselves: their persons, their conversation, and their writings. Ricci signed himself in one work 'Man of the Mountain of the Great Western Region'; in another, 'Man of Paradox'. Crowds came to see them. Sources speak of the house 'besieged', of crowds that would part only when a high-ranking Chinese visitor arrived. Indeed, visitors with personal introductions could number twenty on an ordinary day, a hundred on public holidays. Mandarins flocked to Ricci 'like madmen', and he would spend most of his waking hours receiving them. Some days he hardly had time to eat; the mission's effective dependence on patronage meant that visitors could hardly be turned away. Paying return calls and attending banquets – at which he would be expected to discuss scientific and religious questions, sometimes with learned Confucian or Buddhist scholars – took up yet more of Ricci's time.

Chinese visitors came to see, to wonder, to talk, to learn, to argue. Curiosity about the far West could lead to discussions about philosophy and religion, and the Jesuits were skilled at turning the talk to their own 'way' and its Lord of Heaven. The image of the West and of Christianity presented in microcosm in their house led naturally to curiosity and – they hoped – to attraction to what it had to offer. By 1606, 1,000 Chinese had been baptised; by 1610, 2,500; by 1615, twice as many again. If theirs was an indirect way to make converts, it certainly seemed to achieve results.

Xu was a visitor among the many, but he became a most special one. After passing his *jinshi* exam he was assigned to the Hanlin Academy in the capital for three years, prior to an appointment. Christianity evidently met his needs on several levels, and Western learning held a great appeal for him because of what he perceived as its twin virtues of certainty and practicality. He moved into a house next door to the Jesuit complex, which had a connecting passage into the Jesuit house, and he became a close collaborator and friend of Ricci. The Jesuit taught him mathematics, and soon Xu was asking for scientific books to be translated into Chinese. Ricci insisted that the place to start was the *Elements* of Euclid.

> At my university, all the books of the many different branches of mathematics take this book as their starting point. Every principle, every theory that is being developed, cites this book as its proof.

As soon as Xu grasped the power and precision of the book he became enthused and 'could speak of nothing else with his friends'.

Ricci had a copy of Clavius' *Elements*, in the 1574 first edition, among the collection of artefacts he had brought with him (a note in Clavius' preface explained, indeed, that the book was printed in two small volumes so that it could more easily be carried from place to place; the missions may well have been in his mind). He had tried to make a Chinese version before, but the work, done with another Chinese convert, had stalled after the first book. Next, a friend of Xu's worked with another of the Jesuits: but when this attempt too seemed in danger of foundering, Xu and Ricci themselves took the project over.

It was thirty years since Ricci had studied mathematics. The pair began work in August 1606, Ricci explaining in his Chinese what the Latin text said and Xu writing it down in *his* – probably rather different – Chinese. It was a strategy of translation that, in China, was time-honoured as the method by which Buddhist texts had been translated in the third and fourth centuries. In the Euclidean story it was similar to the methods by which translators from Arabic to Latin seem to have worked in the twelfth century. The two-person method, which he would employ again, enormously increased Ricci's ability to produce texts in Chinese for publication, and it made maximum use both of his understanding of a text that was now embedded in the European scientific tradition and of Xu's knowledge of Chinese mathematical terms and facility with that language.

There were struggles, as there were with any translation. Ricci complained that 'the grammars of East and West vastly differ, and the meaning of words corresponds in a vague and incomplete manner'. He and Xu carefully exploited classical Chinese in order to produce a version that would convey convincingly the style and structure of Euclid's *Elements*. The theorem-and-proof manner was of course retained; so were the initial axioms and definitions, in Clavius' distinctive presentation of them. But it was hard even to find words for 'definition', 'proof' and 'axiom'; as historian Peter Engelfriet explains,

> the word for 'definition', *jieshuo*, is a rather awkward neologism, based on the etymological root of the Latin word; the character for 'proof', *lun*, means not much more than 'discussion', without any suggestion of a compelling discourse; the expression for 'axiom', *gonglun*, means 'public opinion', which does not really render the sense of either a binding statement or a self-evident truth.

The translators, indeed, added something to the structure of the propositions in an attempt to make it clearer; in each proposition they labelled, as well as the enunciation, a *fa* (method) and a *lun* (proof). Nothing in Clavius' edition corresponded to this, and it perhaps arose from Ricci's oral teaching of geometry to Chinese students; it recalled Proclus' attempt to impose a six-part structure on the propositions for the help of his students centuries before. In places, there were further explanatory comments by the translators. Like every other translation and edition, the Chinese *Elements* was a product constructed both from Euclid's text and from other sources: commentary, discussion and explanation.

Apparently Xu and Ricci revised the text three times before they were happy with it. They stopped after the first six books. Six-book versions of the *Elements* were common in Europe for teaching purposes, and although Xu wanted to continue, Ricci was keen to see how this part would be received before translating any more. The work was finished in April 1607, and printed on the Jesuits' own press shortly after.

Each man signed a preface to the Chinese *Elements*, and there they revealed much about their complex motives and the various things they thought the book was and could do. Ricci emphasised the status of the *Elements* in education and its role as a foundation of mathematics: 'everything is contained in his theory and there is nothing that does not follow from it'; 'every principle, every theory that is being developed, cites this book as a proof'; 'anyone who devotes himself to the study of mathematics should use this work as a "ladder"'. Xu took up this theme, and emphasised the systematic quality of the work and the certainty conferred by the deductive method: 'Starting from what is clearly perceptible, [it] penetrates into what is most subtle; from what is doubtful certainty is obtained.' In a separate essay about the *Elements* he wrote that its methods left 'no need to doubt, no need to guess, no need to test and no need to change'. 'One cannot elude it; one cannot argue against it; one cannot simplify it; and one cannot try to change its order.' Echoing some of what Clavius said in his own preface to the *Elements*, Xu argued that the study of Euclid would improve the mind: 'it will train people's inborn intelligence and render it refined and precise'.

Both men also had much to say about the practical uses to which mathematics could be put: from digging wells to curing diseases, from arranging orders of battle to astronomical prediction. But for both, a key point was that the study of geometry could also lead away from itself in the direction of theology. Xu hoped that it would show the teachings of his master were trustworthy and beyond doubt; if the Jesuits spoke the truth about visible things and possessed methods for achieving certainty in the natural domain, surely they must be right about transcendent matters too. Ricci, on the other hand, appeared to think of the *Elements* as a way to introduce to Chinese readers a model of right reason that he could then deploy in religious argument; in his diaries he suggested the *Elements* could function as a crash course in logic.

The book was printed in fairly small numbers, and copies were sent out as gifts; they can be traced in the catalogues of several private libraries from the period. Ricci also sent copies back to Europe, including to Clavius himself. A corrected second edition was printed in 1611, and its text was incorporated in the *Tianxue Chuhan* in 1626, Ricci's 'collection of heavenly learning'. Like many of the artefacts in Ricci's house, the Chinese *Elements* may have been admired more often than it was understood. But several of the Chinese terms Xu and Ricci created, including the use of *jihe* to mean 'geometry', live on to this day.

Xu continued to write about mathematics in Chinese. He began a search of the surviving works of classical Chinese mathematics and attempted to confront these works with the Euclidean methods he had learned. Within a few years he had produced two texts about surveying, the second of which explicitly compared Western and Eastern methods. His study of Chinese arithmetical methods, the *Gougu yi*, attempted to use Euclidean propositions to prove the Chinese algorithms were correct. Xu also believed that the influx of Western mathematics provided an opportunity to repair gaps arising from the incomplete transmission of ancient Chinese texts; but the differences of method proved too great for his attempted synthesis to be satisfactory. In the longer term, Chinese mathematicians would prove willing to adopt Western calculation techniques while largely rejecting the modes of reasoning that underlay them. Subsequent generations of Jesuit missionaries later in the seventeenth century would tacitly accept the failure of Ricci's Euclidean project, translating into Chinese not further Euclidean works but less theory-laden practical textbooks.

Xu Guangqi broadened his interest in and patronage of Western learning, finding new ways to relate it to the practical studies which were so dear to him. He collaborated on works on hydraulics and he pressed successfully for the use of Western learning to reform military affairs. (Ironically, the invading Manchu, who in 1644 occupied China, would do so with the aid of Western arms.) Near the end of his life he was promoted as minister of rites and chancellor of the Hanlin Academy. He continued to write, and took charge of a reform of the calendar based on Western astronomical methods: a strange echo of the work of Clavius in the previous century, half a world away. In all this, Xu was supported both by his own students and by a new generation of Jesuit missionaries. At the time of his death in 1633 he was at work on a treatise on agriculture.

The collaboration of Xu Guangqi and Matteo Ricci was a remarkable moment when personal contact became cultural contact, and it was one of the most extraordinary of all encounters with Euclid's *Elements*. The period of their collaboration effectively ended when Xu returned to Shanghai in 1608 for the period of mourning and funeral for his father.

For Ricci himself the Euclidean project was just one part of an enormous round of writing, translating – including further works on mathematics – printing, and of course personal contact. He described himself criss-crossing the city, riding through the streets with a black veil over his face against the dust. His learning, his memory and his courtesy continued to impress many of those he met.

But it was exacting a toll, and Ricci was no longer young. The year 1610 saw another round of *jinshi* examinations in the capital, and 5,000 or 6,000 hopefuls arrived. As well as their exams they toured the cultural sites, of which the Jesuit church and house were one. For every visitor Ricci had to act as a guide, explaining the sacred objects and images in the church; with each, he engaged in polite and learned conversation, showcasing Western learning, Christian teaching, and European books.

He became ill, and he died on 11 May. By an exceptional imperial favour, instigated by one of Xu's students, a plot of ground was granted for his burial, which would be permanently owned by the Jesuit order. According to a Chinese biographer at the time, the favour was granted chiefly because of Ricci's translation of Euclid.

Blame not our author

Geometry on stage

LINE: Ho ho ho! I see if Carnival continue I shall change my shape and be I know not what. I find that macaroni adds some latitude to my longitude: I think I shall make the scholars wrangle about my definition, for though I am Line, yet I am not all together *lacking breadth*.

These lines were heard by a group of English students in Rome in – probably – the carnival season of 1635, when they witnessed the performance of one of the most endearing – and intriguing – responses to Euclid's *Elements* in the seventeenth century. The play, written in English and entitled 'Blame not our author', was the work of an unknown hand; it took for its main characters four geometrical shapes – 'Quadro', 'Rectangulum', 'Line', and 'Circulus' – together with their 'Ruler' and 'Compass'. There were bit parts for 'Semicirculus', 'Rhombus' and 'Triangulum'.

It may sound a somewhat flat premise. But the plot developed a surprising degree of complexity, because the characters did not live up to what might have been expected of geometrical shapes in the way of purity and stability. They interacted; they suffered. Though flat, they displayed 'depths' of feeling (it was a play full of bad puns: 'compass me', 'out of your element', and so on).

In the play, it was possible for shapes to rebel against the Compass and Ruler who had made them: for them, indeed, to change their shapes. Squares could dream of becoming circles; Rectangulum imagined a sort of cosmic coup led by himself and his four-cornered brethren, which would make him 'Ring leader to the Planetts'. Rectangular planetary orbits? A recipe for chaos, surely: 'a fraction in the universal order of things', as another character put it.

The main plot turned on the desire of Rectangulum's master, Quadro, to become circular. He imagined not so much a geometrical construction that would achieve this (to 'square the circle' had been a notorious geometrical conundrum since antiquity) as a magical metamorphosis in the style of Ovid. Things rapidly became complicated. A pharmaceutical treatment failed to produce the desired result; next, the characters tried out an instrument called 'the Squarenighers daughter', binding Quadro in hoops to make him rounder. This didn't work either, and turned out to have been a malicious joke by Circulus. Once he escaped his bonds, Quadro spent the rest of the play planning his revenge.

The plot thickened further, as Quadro's henchman Rectangulum tried to manipulate the new situation for his own ends. Telling lies all around, he persuaded Triangulum that Circulus had killed Compass. But in the end his lies were exposed and matters were resolved, and the ruler, Regulus, intervened to punish Rectangulum:

> Come Joy, and with thy peaceful olive wreath
> Circle our heads and crown our cloudy fronts . . .
> Let him that squares from rule and compass be
> Vassal to fear and base servility.

'Blame not our author' was never published; it exists in a single manuscript in the English College in Rome, the training college for English Jesuits since 1579. The unknown author was presumably one of the masters there, but the handwriting has not been identified. Perhaps he was the master of the physics class, where the *Elements* was studied by second-year students.

Though it seems to have been practically the only play of its period with geometrical shapes for its characters, 'Blame not our author' had cousins in other more or less whimsical morality plays from various academic locations around the turn of the sixteenth century. They dealt with subjects like the war between the noun and the verb, or 'the Combat of the Tongue and the Five Senses'. Wide-ranging, allusive plays, they were crammed with references to myths, fables, Bible stories, plant lore, ancient history and philosophy: everything needed to flatter the intellects of a student audience. Another almost compulsive fondness was for images of food, providing a hint of what was on students' minds at the English College and elsewhere. If Quadro were to get his way,

> Why now all our square trenchers will be turned into round dishes, our caps quadrangular into the flat caps of the Citizens of London, our fournooked Pasties into *pastichios*.

And, when things were hotting up, Line threatened Circulus:

> let me alone with him: I'll slice him through the midst and make as many triangulars of him as the Cook doth when he sliceth an [egg] flapjack into portions.

As well as providing the characters, Euclid's *Elements* made a direct appearance in the play, in the form of quotations of some its definitions. They were quoted, in fact, from the ubiquitous edition of the *Elements* by Christoph Clavius.

LINE: Where art thou, Rectangulum?

RECTANGULUM: Just perpendicular over your back. [in Latin] *When a straight line stands upon a straight line . . .*

LINE: Are you making definitions upon my back? [in Latin] *A line is straight which is the shortest between two points.* I will go the briefest way to work.

And, sure enough, the 1591 edition of Clavius' *Elements* is still to be found in the library at the English College.

'Blame not our author' brought geometrical shapes to centre stage and used them to create novel, startling, disorienting effects. Melancholic Quadro, with his yearning for an impossible transformation into a circle, showed that human emotion and desire were not always a good fit for the rigid laws of geometry. He and his fellow characters insistently commented, indeed, on the tension between mathematical ideals and reality. For as well as being a metaphor for truth and certainty, geometry could also be adopted for practical uses; uses which sometimes seemed uncomfortable to readers whose experience of geometry was an experience of Euclid's beautiful deductive structure.

CIRCULUS: Why should I subjugate myself to one
Ignoble, slavish, in the hands of all?
Base compass, minion of each pedlar's pack,
Turned and tossed by each Carpenter:
Each Petifogger binds his straying legs
To compass distance for his scribbling.

Meanwhile, Quadro's square-peg experience of life may have had a good deal of truth for the students at the English College in

Rome, divided from their countrymen by religion and divided from their neighbours by language. Geometrical language provided Quadro's (and the author's) way of expressing that disorientation, in defiance of any idea that mathematics should be about order, reason, and law.

> Yet how can my passions stint
> To see the heavens all orbicular,
> The planets, stars and hierarchies above
> All circular – and Quadro, quadro still?

Thus in 'Blame not our author', geometry became a way to express and explore irrationality: dislocation, disorientation and impossible desire. Far from being an improvement for the mind, for poor Quadro geometrical manipulation served as a spur to frenzy and madness. In him 'the "right" angled [was] anything but normal', as one modern commentator has put it; and indeed Compass and Regulus were themselves aberrant, capricious, vagrant. Perhaps, one character suggested, the problems lay with Euclid himself:

> COMPASS: He that most pestered me was Euclid, about a certaine demonstration to be made of a [in Latin] *circle-heptagon*. I assure you he pushed me.

Or perhaps, elsewhere, it was the playwright's 'mathematique brain' which was the 'author of our grief'. There was perhaps a glance here at the proliferation of versions and interpretations of Euclid. By their very number these were now, maybe, making the experience of reading or studying Euclid an experience of confusion and instability instead of one of beauty and perfection.

'Blame not our author' illustrates the complexity that ideas about mathematics and geometry could attain in the seventeenth century. It also illustrates that Euclidean geometry was becoming something to play with and to laugh with: something that could be made to take on new meanings and enact new scenes and dramas. A mantle, indeed, which the ambitious author might don for many different purposes. If geometrical shapes and Euclidean definitions could take part in such a carnivalesque romp as 'Blame not our author', it seems that 'Euclid' was becoming a costume in which many different roles might be played.

Baruch Spinoza
The geometrical manner

Amsterdam, 1656. A Jewish youth is expelled from the synagogue. A small, slight man, aged twenty-three, with a beautiful face, long black hair and a moustache of the same colour, and black eyes. His destiny would change the history of philosophy.

> Cursed be he by day and cursed be he by night; cursed be he when he lies down, and cursed be he when he rises up; cursed be he when he goes out, and cursed be he when he comes in . . . Nobody should communicate with him orally or in writing, or show him any favour, or stay with him under the same roof, or come within four ells [about nine feet] of him, or read anything composed or written by him.

Seventeenth-century Europe saw not a few people use Euclid's *Elements* as a specimen of how thought should be done, how knowledge was structured or how reason really worked. There were titles such as *The Euclid of Logic, The Euclid of Medicine,*

Elements of Jurisprudence and *Elements of Theology*, and many more. There were, of course, good ancient and medieval precedents for many of these, including the *Elements of Theology* by Proclus.

British philosopher Thomas Hobbes, who yielded to none in his love for Euclidean geometry, famously recounted his introduction to the subject:

> In a certain library I saw by chance Euclid's *Elements*, which happened to be open to the 47th proposition of the first book. When I read these words: 'In right-angled triangles, the square on the side subtending the right angle is equal to the squares on the sides containing the right angle,' I immediately said that even if this were true, it could not be known by man, ignorant as I then was of mathematical matters. But upon inspecting the demonstration I was at once sent back to Proposition 46 and from there to others, until I arrived at first principles.

Euclid's *Elements*, in this story, was a model not just for certainty but for clarity and transparency. The logical links were easy to see and easy to check. If only every subject could be like that. Hobbes went on to call one of his major works the 'Elements' of philosophy. He believed that 'the Geometers have managed their province outstandingly', and that if the laws of human action were understood with geometrical certainty, then war, ambition and greed could effectively be abolished.

There was a decided current that took Euclid, then, as simply a replacement for logic as traditionally conceived: Aristotle is dead, long live Euclid. 'Geometry is an excellent logic', Bishop Berkeley would write in 1734 (the context was his defence, as he saw it, of mathematics against dilution by logically problematic novelties). Euclid remained a proverbial standard of certainty through the seventeenth century and into the eighteenth. His theorem-and-proof style acquired a prestige, a reputation for having the simplest possible relationship with the truth, that other ways of writing could never match. Setting aside some of the debates about the status of mathematics and the certainty of its proofs that had been the background to Clavius' work, quite a number of seventeenth-century writers simply wrote their philosophy, physics, or theology in a geometrical style, evidently convinced that the advantages of clarity, certainty, succinctness, transparency and persuasiveness would follow.

One of the several pinnacles in this mountain range was the work of Baruch Spinoza. Born to Iberian Jewish parents in Amsterdam in 1632 and given a name meaning 'blessed', he studied Hebrew, the Hebrew Bible and rabbinic literature at the local Talmud Torah school. He left at a young age to work in the family's business, importing and selling tropical fruit from a stall by the main canal. But by his twenties he had decided that his future was as a philosopher, and he was attracting attention for the unorthodoxy of his opinions. In the summer of 1656, by now openly rejecting the fundamentals of Judaism, he was put out of the synagogue with a blistering writ of anathema.

For most of the rest of his life Spinoza worked as a lens-grinder, making telescopes and microscopes in addition to separate lenses. He moved around the Netherlands, living near Leiden, then near The Hague. He studied Latin, and ancient and modern philosophy: particularly that of Descartes, which he discussed with a circle of friends. He was visited by scientists and philosophers from England and Germany. And he continued to philosophise.

For a student named Caesarius he prepared notes on parts of Descartes' *Principles of Philosophy*, in a Euclidean style. His account of the matter had axioms, definitions, propositions and proofs, and demonstrated in a geometric manner some of the main

Baruch Spinoza.

conclusions from Descartes' book. His friends prevailed on him to share the text, and his two-part exposition of the 'principles' became his first published work in 1663: and in fact the last work he would publish under his own name.

Evidently mathematics, geometry and Euclid had special importance for Spinoza. He owned mathematical books by ancient and modern writers. He had already used a geometrical style to set out proofs of the existence of God, both in a letter to Henry Oldenburg, secretary of the Royal Society in London, and in an appendix to an unpublished short treatise on philosophical topics. By rearranging Descartes' proofs in the same style he was going further, though still for a basically educational purpose, and indeed acting directly against Descartes' own opinion that the geometrical method did not 'engage the minds of those who are eager to learn, since it does not show how the thing in question was discovered' (Descartes had even provided a specimen – four propositions and a corollary from his philosophical *Meditations* – to demonstrate this). For Spinoza, there was evidently no such objection. The preface to his version of Descartes' *Principles* stated emphatically that both for finding the truth and for teaching it, the best way was 'that of the Mathematicians, who demonstrate their Conclusions from Definitions, Postulates, and Axioms'.

On the other hand he was not convinced, unlike many of his contemporaries, that applying mathematics to the natural world was a way to make progress. For Spinoza empirical evidence – even the most quantitative – was always suspect and could be of at best something like heuristic value, as it could never lead to certain knowledge.

After his Cartesian publication Spinoza worked, or continued to work, on an exposition of his own philosophical ideas – which were distinctly different from Descartes' – for his circle of learned friends. At one stage he experimented with writing systematically in a non-geometrical, discursive manner, but by the early 1660s he was reworking his material into the geometrical style, which he had now decided was the best – if not the only – way to present it. He continued to draft and send sections to friends for comment over several years.

Spinoza paused in mid-1665 to write a (non-geometrical) theological and political treatise. Its stark reworking of traditional concepts such as God and scripture made it almost unpublishable, and it was issued anonymously in 1670. It made him famous across Europe – its authorship was quickly an open secret – but the political situation in the Netherlands was deteriorating, with the notorious murder of the republican de Witt brothers by a monarchist-inspired mob in 1672, and two years later the treatise was banned by both the Reformed Church Council of Amsterdam and the (secular) Court of Holland.

Spinoza returned to his philosophical work, to be known as the *Ethics*, knowing now that publication would be difficult if not impossible. He refused offers to move to Paris or Heidelberg, remaining in the troubled Netherlands to finish his book in the first half of the 1670s.

It was a remarkable piece of work, one of the boldest philosophical treatises ever composed. It amounted to a sustained attempt to work out the consequences of the nature of God as Spinoza conceived it, for every area of thought: minds, morals, emotions, politics, religion. It was, as completed, a superficially most austere work: on opening it, the reader was faced immediately with definitions, with no preface or attempt at guidance or introduction. The book was in a geometrical style throughout, from definitions (substance, attribute, mode, God, eternity) and axioms ('the knowledge of an effect depends on and involves the knowledge of a cause') through 207 propositions, each with its proof.

The austere style did something to obscure the work's sources: many of the ideas and a few of the conclusions were from Descartes and Hobbes, or from the Jewish philosophical tradition including Maimonides and ben Gershom. But on the whole it was strikingly, indeed radically original. Spinoza's answer to the most basic philosophical question of all – what *is*? – was simplicity itself: God. For him, everything else existed not independently but as a modification, a 'mode' of God. Thus everything humans could possibly want to know was implied, enfolded, in the nature and properties of the divinity. 'Modes' had attributes: specifically, physical things had the attribute of extension, while minds had the attribute of thought (and the two did not, strictly speaking, interact).

There is some question why Spinoza used the term 'God' rather than 'nature' (or something else) for his basic entity; certainly there was no question of Spinoza's God doing such things as loving human beings, or indeed of being holy or sacred or desiring to be worshipped. Nothing, indeed, could have been more impersonal than Spinoza's God, and in his disdain for anthropomorphism there was probably a glimpse of the views (and the obnoxiousness?) that got him expelled from the Amsterdam synagogue: 'if triangles could speak, they would assure us that God is eminently triangular, and if circles could speak, they would assure us that God is circular'.

Chance, free will, and cause and effect as most people understood them also fell away in the *Ethics*. If everything was the logical consequence of the divine nature, there was no room for choice or contingency:

> from God's supreme power, or infinite nature, infinitely many things in infinitely many modes, i.e. all things, have necessarily flowed, or always follow, by the same necessity and in the same way as from the nature of a triangle it follows, from eternity and to eternity, that its three angles are equal to two right angles.

So, 'things could have been produced by God in no other way, and in no other order than they have been produced'. There were, so to speak, reasons for everything, but causes for nothing.

This was a bleak world. Spinoza's view was that people could not control what happens, nor could they control how they reacted to it. They could not even control their emotions. He went beyond the Stoics and their celebrated doctrine of resignation, demanding that humans should accept even that they possessed no freedom and no choices. One of the few crumbs of comfort was in his echo of a doctrine also adopted by ben Gershom, as it happens: that that part of a mind which consists of true ideas was by that very fact eternal and did not die with the body.

So when Spinoza's *Ethics* eventually, in its final part, turned to the subject of ethics, there was in a sense not a great deal to say. He found no room for moral responsibility, and virtue or excellence consisted for him merely of an innate tendency for biological entities to seek their own preservation and their own advantage. For human beings, virtue and advantage – and immortality – meant simply knowledge. The goal of life was to replace false beliefs with true ones, to rise above the buffeting of the emotions, and to achieve a sort of infinite resignation, acquiescence in the necessity of all things.

The *Ethics* was an astonishing performance, and it received the widest possible range of reactions. It was eventually printed a few months after Spinoza's death in 1677, initially in the Latin in which he had composed it, but soon afterwards in a Dutch translation. It became an enormously influential work, setting the agenda for much that would subsequently be called rationalism among both his followers and his opponents. To this day, almost every aspect of the book's interpretation is contested by philosophers, with basic disagreement about the most fundamental questions of what Spinoza meant and what it implies. The geometrical style in which he wrote has been admired for its clarity and precision, but equally often derided for its pretentious irrelevance. Friedrich Nietzsche, who agreed with Spinoza about many fundamentals, referred derisively to the 'hocus pocus of mathematical form' of the *Ethics*. Most modern discussions seem to concur in using words like 'forbidding', 'intimidating' and just plain 'difficult' to characterise it. But Spinoza surely meant to create clarity and precision by adopting a form with which many readers in his century would have been familiar, and which they would not have found forbidding or intimidating.

Spinoza stood in a direct line with those of his own and previous centuries who believed geometry had a special relationship with truth, and geometrical method a special status as a result. Proclus would have recognised his methods and ben Gershom some of his conclusions; Matteo Ricci would surely have concurred that knowledge, essentially, is shaped like Euclid's *Elements*.

That was the key, really, the unstated axiom that Spinoza assumed everywhere: that knowledge, real knowledge, worked like geometrical deduction; that it was structured as a network of logical relationships; that it consisted of the logical procession of consequences from assumptions. And so, that 'the order and connection of ideas is the same as the order and connection of things'.

This was a clue to what could otherwise seem Spinoza's arbitrary project to present his ideas in geometrical form. Yes, that form had real advantages as a rhetorical style: it was clear, precise and concise, particularly in its ability to refer quickly to several previous results when proving a new one. It was persuasive, in the sense that unless an interlocutor could find an error in the reasoning, assent was required: as Descartes put it, 'the reader, however argumentative or stubborn he may be, [was] compelled to give his assent'. And, since it was made up of short sections, it could be interrupted as needed with 'commentary' to hammer points home using more traditional rhetorical tools and appeals to the emotions such as irony and sarcasm.

But much more than that, the geometrical arrangement was the right one for Spinoza's *Ethics* because it mirrored the structure of things that the book described. The book started with a few premises and deduced everything logically, just as Spinoza's universe started with God – and the properties of God – and everything else followed logically. The pattern of endless deduction presented in the book really was, for Spinoza, how the world, the nature of things, was structured. Rigorous deduction mirrored the necessity that ruled in nature. Arguably, to use a geometrical method of presentation was the only way for Spinoza to display this feature that was so central to how he saw the world.

Finally, the geometrical method, as well as persuading readers and displaying the structure of reality, also had in Spinoza's opinion the capacity to improve readers' minds. It could inculcate the special virtue not just of knowing the right things but of knowing them in the right way. It could make the order and connection of ideas in people's minds mirror the order and connection of things in reality. For Spinoza, this was close to the basic task of ethics: cleansing the mind of wrong ideas and replacing them with right ones; countering emotion and turning chatterers into reasoners. To work through geometric-style proofs was, he intended, a kind of self-improving therapy for readers of his *Ethics*.

In all of these ways Spinoza – who had very largely repudiated the philosophical tradition that preceded him – was paradoxically close to some of his predecessors, particularly the Platonists for whom everything emanated from the One and for whom all knowledge was to be deduced from the properties of the One. Spinoza's God was not the Platonist One; but like them, he had found a route to true knowledge and true human improvement that led inexorably through Euclidean geometry.

Anne Lister

Improving the mind

Halifax, Yorkshire; Tuesday 13 May 1817. A young woman records her day in her diary.

> Between 1 & 2, the 1st 7 propositions of the 1st book of Euclid, with which I mean to renew my acquaintance & to proceed diligently in the hope that, if I live, I may some time attain a tolerable proficiency in mathematical studies.

The belief that true knowledge was structured like Euclidean geometry peaked with Spinoza. After him, few major thinkers, if any, used the 'geometrical manner' at such length outside geometry itself. The philosophical world was changing – Spinoza had done a good deal to change it – and the assumptions on which thinkers from Plato to ben Gershom had founded the cosmic importance of geometry were no longer viable. Eternal archetypal forms were out of fashion, and with them geometry's exalted status as a bridge from the mundane to the divine, from the human mind to the mind of God.

But the belief that studying geometry was good for you lingered: and did so to quite an astonishing degree, even once its philosophical foundation had gone. John Locke, surely Britain's favourite theorist of the mind throughout the eighteenth century, had stated that mathematics was 'a way to settle in the Mind an habit of Reasoning closely and in train'; that 'in all sorts of Reasoning, every single Argument should be managed as a Mathematical Demonstration'. As with Spinoza and many who had preceded him, knowledge was for Locke a kind of moral virtue: not just knowing the right things but knowing in the right way. Geometry – it seemed to many who echoed Locke's sentiments, in Britain and beyond – was the right way.

A little story from the sixth book of Vitruvius' *Architecture*, from the first century BC, became popular in the seventeenth and eighteenth centuries as an illustration of geometry's identification with humanity, reason, and the well-functioning, civilised mind. In the story Aristippus, a Socratic philosopher, was shipwrecked and thrown upon the beach at Rhodes. He saw some geometrical diagrams drawn there, presumably in the sand, and exclaimed to his companions, 'Take heart, for I see the marks of humankind.' Apes, evidently, do not draw geometrical diagrams in the sand, and neither do barbarians. The story was a resonant one: David Gregory's 1703 edition of Euclid's works featured a picture of Aristippus on the beach; so did Edmond Halley's 1710 edition of Apollonius and, at the end of the century, Giuseppe Torelli's 1792 edition of Archimedes.

Euclid's status was now partly a commercial matter, since publishers had every reason to praise and promote this longest-running of bestsellers. The market for new versions of the *Elements* showed no sign of slowing in the eighteenth century, and editors and publishers were ingenious in finding ever new ways to present the text. They printed it in new languages (Swedish in 1744, Hebrew in 1775, Portuguese in 1792). They added the fashionable, and increasingly widely understood, algebraic proofs in place of geometric ones. They larded the text with new, fuller and better explanatory comments than ever before.

It was in England and the English-speaking world that the mania for studying Euclid burned hottest for longest. A relatively conservative mathematical tradition reigned there at the level of research and university teaching, and a thorough grounding in Euclid was one of its supports. By the end of the eighteenth century, England alone had seen the publication of nearly fifty works which were in whole or in part versions of Euclid's *Elements*.

Euclid's continuing visibility was, indeed, also to some degree an institutional matter: the *Elements* was named on the curriculum at a number of influential universities. In the mid-eighteenth century David Gregory – son of the mathematician, and first Regius Professor of History at Oxford University – made Euclid one of the requirements in the examinations at his college. Euclid was being taught at Harvard by the 1730s. Other institutions were effectively obliged to follow their lead.

The schools that fed the universities naturally tried to prepare their students with a preliminary study of the *Elements*, or some of it, with the result that – particularly in the English-speaking world – Euclid achieved an increasing dominance over mathematical education across grammar schools, private academies and private tuition. That dominance was also, partly, a social matter: the fact that elite universities and schools thought Euclid important led any number of people to conclude that one of the things they must do in order to become properly educated was to study Euclid. Self-teaching and mutual teaching at local mathematical clubs flourished, and many were the individuals who read and reread Euclid's *Elements* in the belief that it was doing them some social or intellectual good, or both.

Lady Anne Conway, in London in the 1650s, taught herself arithmetic and engaged a tutor to get her through the *Elements* when she found she could not manage the book alone. She gained a reputation as a 'quick proficient' in the subject and received a letter from her brother urging her to limit herself to the first six books, 'For the study of the Mathematiques may take up all ones life.' Her husband was moved to take up Euclidean study as well, despite his engagement in affairs of state (this was in 1657–8, perhaps the most tricky period of the English Commonwealth). A few years later, his former tutor Henry More, well known as a philosopher in the Platonist tradition, retreated to Cambridge during an illness and became convinced that the study of geometry as a respite from theology had – together with the change of air – cured him of his fever: 'I had almost forgott all that little I knew in Geomettry [but] which I have recovered with some advantage, and my health, I hope, into the bargain.'

Later, through the eighteenth century, Britons with mathematical aspirations had a ready shop window for their talents in the form of the mathematical magazines: annual or sometimes quarterly or monthly question-and-answer publications that printed selections of mathematical problems and readers' solutions to them. The *Ladies' Diary* and the *Gentleman's Diary* were probably the best known, and nearly half of their mathematical problems were geometrical, many of them solved by purely Euclidean methods and some citing Euclid explicitly. This was a way to make public – albeit often behind a pseudonym – Euclidean attainments that would otherwise have remained wholly private: often the attainments of those who because of their class or their sex missed out on formal education in the subject and had educated themselves with the help of a book or a tutor. Typical was one Alex Rowe who in 1797 asked how to draw a right-angled triangle given only the lengths of the two lines that cut it in two through its other (non-right) angles (bisecting them, to be specific): an elegant puzzle that could easily have appeared in the *Elements*, although in fact it didn't. About two dozen readers gave correct solutions.

Anne Lister was the oldest surviving child of a landowning family fallen on (relatively) hard times, her father an army captain, her inheritance Shibden Hall in Halifax, Yorkshire. She is a gift to the historian because of the twenty-seven volumes of her diary, totalling almost 4 million words and mostly not yet thoroughly studied or published (about a sixth are in code).

She attended private schools in Ripon and York and at the age of sixteen she 'began Euclid' while attending the school of a Mr Knight in Halifax. Allowing for some holidays, she seems to have worked through about one book of the *Elements* each month, finishing book 5 in August and book 6 in September, when she apparently stopped. Mr Knight became the vicar of Halifax and Anne Lister became the de facto manager and guardian of the Shibden estate; she proved to be an astute businesswoman with a good head for figures.

Unconventional in more than one respect, she adopted masculine clothes of severe black and styled herself a 'gentleman', exercising oversight of the household and estate with the assistance of a steward. Anne Lister has been called 'the first modern lesbian'. Anxious about rank and her father's behaviour which she reckoned crude, coarse and 'unlike a gentleman', she developed a conservative, starchy public persona including a paradoxical opposition to classical education for women, that her own behaviour defied. Locals called her 'gentleman Jack' and were on occasion very unkind about her peculiarities.

Anne Lister.

In determined – and successful – pursuit of the knowledge and mental attainments of university-educated gentlemen, she put herself through a programme of reading that ranged from Sophocles, Ovid and Juvenal to Malthus, Rousseau and Walter Scott. And, naturally, Euclid. On 13 May 1817, now aged twenty-six, Anne took up the *Elements*, apparently for the first time in nine years. She worked through the first seven propositions, from the construction of an equilateral triangle (proposition 1) to the proof that only one triangle can be made from a given set of three lines (proposition 7). Her plan was to 'proceed diligently' and acquire proficiency in mathematics, of which she evidently considered the *Elements* the foundation. Ultimately her intention was to become an author and 'to turn my attention, eventually & principally, to natural philosophy'. She resolved 'to devote my mornings, before breakfast to

Greek, & afterwards, till dinner, to divide the time equally between Euclid & arithmetic', subsequently tackling algebra. 'The afternoons & evenings are set apart for general reading, for walking, ½ an hour, or ¾, practice on the flute.' She called in Mr Knight's assistance when she felt she needed it.

She reached the end of book 2 by late July and continued with Euclid into the autumn. She was consciously studying more thoroughly than she had at school, struggling successfully, for instance, with proposition 2.13: that, in an acute-angled triangle, the square on the short side is equal to the product of the other two sides *plus* the product of the triangle's base and height (it is far from simple). She recorded that 'I was very stupid about it the last time I did over the 1st six books of Euclid, & tho' I have comprehended it thoroughly this morning, yet it has cost me above ½ hour – it is certainly my *pons asinorum*.' (The *pons asinorum*, the bridge of asses, was traditionally proposition 5 in book 1, which both had a diagram looking vaguely like a bridge and proved impassable for weaker geometrical talents.) By November she could boast to Knight that she had not only worked over books 1 to 6 twice, but also studied thirty propositions of Euclid's other geometrical book the *Data*: more, frankly, than most university graduates.

But family worries, including a series of illnesses and deaths among her relatives, disrupted Lister's studies. Later in life, after she inherited her family's estate, she travelled widely, and she died of what seems to have been an insect bite at Kutaisi, in the country of Georgia. It is not clear that she ever returned to Euclid after 1817: but her determined study in her mid-twenties, evidently desiring to stand in the long line of those who had improved their minds by contact with Euclid, was impressive by any standard.

Interlude

Wait. Stop (again). Several of the characters in this story – about Euclid as a sage, a philosopher – have gestured beyond that particular horizon, to the use of Euclid's *Elements* in the practical parts of life. As Anne Lister herself said, reasons for reading Euclid pointed beyond merely individual mental improvement, and towards natural philosophy and its much more public sphere of action and innovation.

Was Euclid not at least as much master builder as he was learned sage? Had he not, for centuries, been read for practical, public reasons just as much as for private, philosophical ones? Didn't his book affect the world just as much as it affected the individual reader?

Yes, it did, and there is another Euclidean story to tell. This one has its roots in the Egyptian mud, the measuring of land and the calculating of taxes.

III

Hero

Petechonsis

Taxing and overtaxing

A strip of land in the Nile Delta, near Thebes. It is fertile and regularly floods, until the construction of barrages and dams in the nineteenth and twentieth centuries. One spring around the end of the first century BC it is surveyed for tax purposes, a scribe writing down the results: 'The land measurement of the [land] of Kalliedon [farmed] by Petechonsis beginning on the west.'

The survey records the location, dimensions and area of each of the fields. They are not squares or rectangles but irregular shapes; the first one has its sides measured as 15, 28, 6 and 32 thirty-seconds of a schoinion. (A schoinion is one hundred cubits or perhaps around fifty yards, so these thirty-seconds are a few feet long.) The surveyor then works out the area by multiplying the averages of the two pairs of opposite sides, and rounding (up). The answer he finds is ten thirty-seconds of a square schoinon: a little under 800 square yards. A little too large, in fact. Petechonsis will be overtaxed.

Weighing and measuring, surveying and accounting, are virtually as old as writing; in some places they may be older (not every way of recording numbers requires written language). By the end of the fourth millennium BC, certain Mesopotamian city states had powerful written techniques for counting – and thereby controlling – things and people. From the early third millennium BC, field plans survive from Ur, showing the land divided into triangular and rectangular plots whose areas were relatively easy to calculate. The reed and the rope became symbols of the fair measurement of land, and of justice in general. Hebrew culture used the same image: 'Diverse weights and diverse measures are both alike an abomination to the Lord.'

The Egyptians, too, had been measuring the earth for thousands of years by the time the Nile Delta became a Greek province. A famous story was recorded by Greek historian Herodotus in the fifth century BC. King Sesostris, it was said,

> divided the country among all the Egyptians by giving each an equal parcel of land, and made this his source of revenue, assessing the payment of a yearly tax. And any man who was robbed by the river of part of his land could come to Sesostris and declare what had happened; then the king would send men to look into it and calculate the part by which the land was diminished, so that thereafter it should pay in proportion to the tax originally imposed.

This, in Herodotus' opinion, was the origin of surveying among the Egyptians.

The story may have a grain of truth to it. Certainly a preference for land as a basis of taxation led, in Egypt, to a long-lived tradition of surveys: sizes of fields were already being measured in the Old Kingdom period, the second half of the third millennium BC.

From the Middle Kingdom down to the Roman period there are sources showing substantially the same techniques in use. For the Egyptians as for the Babylonians, the correct surveying of land was a symbol for justice. The 'Teaching of Amenemope' exhorted the hearer:

> Do not move the markers on the borders of fields,
> Nor shift the position of the measuring-cord.
> Do not be greedy for a cubit of land,
> Nor encroach on the boundaries of a widow.

Under Greek rule, the Egyptian surveys continued as they always had, the basis for raising (or occasionally lowering) rents and for the preparation of leases and land transfers. The invariable technique was to estimate a quadrangle of land by adding opposite sides, halving the two sums and multiplying them together.

Measuring an Egyptian field.

It was not an accurate method, strictly speaking. If the piece of land was a rectangle it gave the right answer; otherwise it would overestimate: which of course was desirable from the landlord's point of view. The authorities sometimes, indeed, recorded a 'difference of measurement', but the tenant was still required to pay tax on the larger amount. The calculation done this way was quick and easy, though, and its answers reasonably satisfactory. The technique was still reportedly in use in Egypt in the nineteenth century.

Traditions of practical, everyday mathematics, of measurement and calculation, exist around the globe, and certainly existed in the ancient Greek and Egyptian worlds that were the cradle of Euclid's *Elements*. It is natural to wonder how or whether Euclid's learned, theoretical geometry was related to such things.

Over the years there have been some heroic attempts to relate particular parts of the *Elements* to origins in specific practical concerns. None has really achieved consensus – nothing in the *Elements* looks like the division of a real field or the planning of a real city – and it would be a bold historian today who claimed any straightforward development from practical life to the definitions and proofs of Euclid's geometry.

The Greek evidence for Euclidean geometry likewise paints an ambiguous picture of its relationship to practical activities. There are hints of a chasm between the theoretical and the practical. Practitioners had a professional structure, an institutional framework, while theoreticians were learned amateurs. The two mostly ignored each other and indeed were largely invisible – socially speaking – to one another. When they did interact, learned writers sometimes lamented practitioners' use of 'mechanical' methods and constructions instead of proper proofs. Meanwhile playwrights such as Aristophanes were happy to treat theoretical geometers as figures of fun.

Yet clearly the two kinds of geometry had something to do with each other. The name of the theoretical discipline, *geometria*, had originally meant measuring the earth – surveying – and continued to mean that in Hellenistic Greek; a *geometres* was the hard-handed man who actually wielded the rod and rope on site. Euclid's word for a perpendicular, *kathetos*, literally meant 'let fall': like a plumb line. According to one ancient commentator the name *tomeus* for a sector of a circle was suggested by the Greek word for a shoemaker's knife. Even the Greek word for definition, *horos*, was not safe from practical associations: a *horos* was originally a boundary, a landmark.

And the chasm was not absolute. Euclid's near-contemporary Archimedes was a high-calibre theoretical mathematician of whom, also, numerous tales circulated about machines invented and mechanical marvels devised. In one account – that of Xenophon – even Socrates acknowledged that the practical uses of geometry were legitimate, before going on to examine how its higher reaches could help the mind reach the apprehension of the Good.

Thus the relationship between practical and theoretical traditions in mathematics has long been complex, and has often seemed a puzzle to those involved in one tradition or the other. Was mathematics a single thing or several? Ought there to be a clear relationship between Euclidean geometry and its practical cousin? Did either have anything of worth to teach the other? Over the centuries, as both strands have evolved, the relationship has changed many times, and for practitioners and theoreticians alike it has sometimes seemed worthwhile to try to strengthen it: to make theoretical mathematics (more) useful and to make practical mathematics (more) rigorous. Such projects would make Euclid into a kind of Promethean hero, bringing the light of proof and rigorous deduction to the world of practical mathematics, bestowing on mortals true knowledge where before they had only guesses and estimates. On the other hand, such a relationship would also transform the introverted philosophical *Elements*, making its endless tangle of internal references point outwards; making its cold logical structure a bearer of meaning in the real world.

Dividing the monochord

A royal festival at Alexandria, early in the third century BC; perhaps during Euclid's lifetime. Perhaps Euclid is present.

The musician Nicocles of Tarentum is performing: singing to his own accompaniment on the cithara (like a lyre). The repertoire is old but the performer's skill is always new, rendering the complex elusive harmonies from centuries before. A victor at games at Delphi, Corinth and Athens, he leaves his audience enraptured, and bears away the prize once again.

It seems that one of the first people to make practical use of the *Elements*, its results and its methods may have been Euclid himself; though the evidence that has survived makes it difficult to get beyond that 'seems'.

Over the following centuries, many works accumulated around Euclid's name, concerned with various practical topics, and in some cases it is now very hard to be sure who wrote what, or whether certain pieces of text do or do not go back to Euclid himself. A collection of fifty-eight propositions on optics seems to be authentic; it was a slight work compared with the *Elements*, though it gave the impression that the subject was well established. It described how appearances were formed at the eye in terms of the relative size, shape and position of objects. Closer objects made wider angles, as did larger ones, and so on. Everything was translated into geometry, and what could not be so translated (colour, for instance) was ignored. The book was in a thoroughly Euclidean style and depended on the *Elements* repeatedly. It did not quite go as far as a full theory of perspective – in the sense of what shapes should be drawn on a flat surface in order to provide the illusion of seeing solids – but it did some of the preliminary work that would eventually be used to that end.

The *Optics* also reads strangely to a modern reader because Euclid assumed that the eye emitted rays which made perceptible the objects they struck. Since the lengths of these rays were not thought to be perceived, it was not clear how the eye could distinguish a near object from a large one. Perhaps it couldn't: Euclid considered the vision of only one eye at a time, and in a sense what he wrote about were the illusions that result from closing one eye.

Astronomy, too, was early on drawn into the network of relationships that is the *Elements*. Proclus picked out a few propositions that he said were used in astronomy. And, again, there is a text called the *Phaenomena* treating the geometry of the sphere in a long series of propositions that looks to be authentically Euclidean. It studies the rising and setting of stars and the movement of the sun; it attempts to deduce the length of daylight at a given point on the earth's surface.

On a rather different subject, two treatises about music were being attributed to Euclid by later antiquity, one of them containing a substantial passage in the theorem-and-proof style of the *Elements*. Debate continues among specialists as to whether any part of this musical material goes back to Euclid himself. To a modern mind it is perhaps surprising that geometry should be applicable to music at all.

Music was everywhere in Greek culture, and much that later ages would appreciate as poetry was heard by the Greeks as song. Greek tragedy and comedy had more in common with opera than with spoken theatre, and some of the Greeks who are now thought of as poets might better be called composers. Traditional airs were played on wind instruments (*auloi*) during rituals; both epic recitation and choral song were accompanied by the *kithara* (lyre). Rhapsodes, citharodes, aulodes and auletes competed for prizes at festivals not only in classical Athens, but also in Hellenistic cities like Alexandria. The five days of festivities for Alexander's wedding at Susa involved 'a rhapsode, three psilocitharists, two citharodes, two aulodes, five auletes . . . three tragic and three comic actors, and a harpist'. A procession organised by Ptolemy II at Alexandria involved a choir of 600 men accompanied by thirty citharas.

Through the long period of ancient Greek musical performance, styles and tastes changed. Instruments, tunes and scales had their origins in prehistory, but particularly during the fifth century BC the number of notes available or in use increased, and changes of scale within a piece became more frequent. By around 440 BC one of the Athenian comic playwrights had the dishevelled figure of Music complain to Justice of the more and more outrageous treatment she had suffered in recent times.

Music, particularly music that was increasing in complexity, prompted the development of music theory. Musicians presumably tuned by ear, but some theorists wished to be more precise. The now poorly documented Pythagoreans were credited with noticing

Playing the cithara.

that two sounding bodies – strings, say – whose sizes are in the ratio 2:1 will produce the pleasant musical interval of an octave, and similarly with a few other ratios: 3:2 gives a perfect fifth and 4:3 a perfect fourth. Why? One possibility was to say that the phenomenon had something to do with rates of vibration: it is clearly visible that a shorter string vibrates faster, and audible that it makes a higher pitch. But the Greeks had no way to measure vibrational frequencies exactly; the vibrations were too fast to count. And no one had any idea why simple ratios of frequencies should sound pleasant, any more than they knew why strings whose lengths made simple ratios should sound pleasant together. It was a brute fact, and that was all.

Several Greek thinkers worked on the subject of musical ratios, asking questions about the addition of musical intervals to make larger ones and their subdivision into smaller parts, as well as wondering whether some general principle determined which ratios were related to pleasant musical sounds. Their main conclusions were known to both Plato and Aristotle. Plato gave music important roles in both education and the structure of the cosmos, arguing that musical forms such as simple ratios were special in themselves, and musical consonances merely their expression in sound.

The first extended discussion of musical ratios to survive from ancient Greece is a text called *The Division of the Monochord* and traditionally associated with Euclid. Like the *Elements* itself, it brought together into a systematic argument observations and results from previous generations (and stripped from them anything redolent of Pythagorean number-mysticism). As in the works on optics and astronomy attributed to Euclid, its author assumed some results from the *Elements*, in this case matter from the books which dealt with ratios of numbers: how to put a ratio into its lowest terms and what factors the numbers in a ratio can possibly have.

The *Division* started with a discussion of what sound was, alluding to the relationship between pitch and frequency. After that introduction came mathematics. The text provided no definitions, postulates or common notions (though it did make some hidden assumptions); instead it moved straight into a series of propositions about ratios. For example: a ratio of consecutive numbers – 3:2, say, or 4:3 – cannot be divided into two equal ratios without using so-called irrational numbers or 'surds'. Some of this related closely to matter in the arithmetical books of the *Elements*. The text noted that some simple ratios corresponded to pleasant musical intervals, though it provided no very convincing explanation for which ratios those were, and after a discussion of how those intervals related to each other and to the musical scale, it finally culminated by showing how to 'divide the monochord'. That is, how to take a single musical string with a ruler attached beside it, and using geometry divide the ruler so as to show where the frets or the fingers must lie upon the string so as to produce a musical scale.

It is a most intriguing text: both an exercise in Euclidean number theory and an attempt to describe Greek musical practice with mathematical precision. The author's technical dexterity did not quite disguise the fact that the whole project did not really hold together. No observations were available to link ratios of frequencies with ratios of string lengths; no mathematical criterion was available to determine which ratios would correspond to consonant pairs of sounds.

The Division of the Monochord, furthermore, seemed to inhabit and describe a somewhat idealised world, in which sounds were perfectly in tune or they were not, with no room for the 'almost' of actual experience and real musical practice. The author of the *Division* took certain mathematical facts and worked out their consequences for musical tuning; and that was all.

Despite – partly because of – these oddities, the text proved influential, with its alluring attempt to make music a precise, quantitative field of study. Later writers on music theory quoted from it at length, and it became an important source for the science of 'harmonics' – one of the seven liberal arts – as studied in the medieval universities. As late as the seventeenth century there were serious attempts to build and demonstrate musical instruments to this text's mathematical specifications.

Was it really written by Euclid? The ancient tradition certainly agreed that he wrote *something* on the subject, and the earliest manuscripts of the *Division* state that this was it. But many scholars have doubted it, and the matter remains disputed to this day. Even if it is not by Euclid, though, the *Division of the Monochord* remains unmistakably Euclidean, illustrating that at a very early date Euclid's mathematics, and specific propositions from his *Elements*, were being put to musical uses far removed from the purely geometrical world.

Hyginus

Surveying the land

A new Roman colony, around AD 100. Hyginus is the surveyor of the site. He has consulted the new colony's founder and an augur, and together they have chosen the right spot for the division of land to start. He plants his instrument in the ground: called a *groma*, it consists of a horizontal metal cross with a plumb line at the end of each arm, the whole thing mounted on a wooden pole. If the day is windy, he must also set up a windbreak, to keep the plumb lines from swinging about. He aligns it north–south and east–west, and by sighting along it lays out dividing lines in those directions: the *decumanus* and the *cardo*. He measures out the proper distances to make a large square, 2,400 Roman feet on each side. He measures and sights again, dividing the square into a hundred square smallholdings.

Later, the squares will be allocated by lottery, and the next part of Hyginus' job will be to take new settlers to their land in person, to avoid mistakes or disputes. Finally, he will draw up a map on a bronze tablet, as a more or less permanent record of who owns what.

Roman feats of engineering such as roads and aqueducts depended on accurate surveying of the land on which they stood; Rome's acquisition of vast swathes of territory from the second century BC onwards resulted in the surveying and division of huge areas into farm-sized parcels, carving up newly conquered or confiscated land for the benefit of Roman veterans or the landless.

Trial and error, and gradual improvement, were certainly involved in the development of a Roman surveying tradition, and the instruments used were simple ones. The Roman surveyor was adequately equipped who had writing and drawing materials, sundial, plumb line, a rope or measuring rod, and a *groma*. The *groma*, indeed, was so characteristic of the Roman surveyors that they eventually came to be known as *gromatici*. (Exactly how the *groma* was set up is something of a puzzle; the archaeological evidence, consisting mainly of a broken one from Pompeii, is not really consistent with surviving descriptions of the instrument.)

The *groma* enabled straight lines to be laid out, and squares and rectangles. It did not allow angles to be measured; the Roman surveyors did not do trigonometry. For changes of height, which occurred in the design of aqueducts, they had to use a separate instrument called the *libra*. But the *groma* was perfectly sufficient for laying out a flat grid pattern on the land, which was just what the surveyors most often had to do.

The process was called centuriation – dividing the land into hundreds (of farms) – and it boomed from the second century BC in Rome's new colonies. The Roman Republic and the Roman Empire planned land use more carefully and accurately than in any other country at any time up to the late eighteenth century, and its results can still be seen – especially in aerial surveys – over hundreds of square kilometres of northern Italy and North Africa, in the Po Valley, Campania and Tunisia.

When they were not called *gromatici*, the Roman surveyors were called *agrimensores*: field measurers. As well as making initial divisions, they occasionally surveyed already-divided land for taxation purposes; they also measured the sizes of irregular plots and adjudicated in disputes about land boundaries. The surveyors were increasingly respected as professionals, their services valued, and they were trusted with quasi-judicial functions. The Roman magistrate Cassiodorus once recommended the use of a land surveyor to resolve a dispute, instead of weapons.

> The land surveyor is entrusted with the adjudication of a boundary dispute that has arisen, in order to put an end to wanton arguing. He is a judge, at any rate of his own skill; his law court is deserted fields. You might think him crazy, seeing him walk along tortuous paths. If he is looking for evidence among rough woodland and thickets, he does not walk like you or me; he chooses his own way; he explains his statements, puts his learning to the proof, decides disputes by his own footsteps, and like a gigantic river takes areas of countryside from some and gives them to others.

The tradition of Roman land surveying continued over a period of perhaps 900 years, and naturally it produced its share of writing. For all the longevity and reach of the surveying tradition, though, the surveyors were known more as a class than as individuals, and there is practically none for whom the sources give even so much as a name, a century and the province in which he worked. Hyginus is one, for whom there are surviving writings and the report that he worked around AD 100: but it is not clear where. He wrote about how to establish boundaries and how to designate different kinds of land on maps, and about how to resolve land disputes.

At some point in the fourth or fifth century, a collection of texts about surveying, including Hyginus', was compiled together by an unknown editor, perhaps based in Rome, who removed introductions and conclusions, paraphrased and compressed, but made little effort to create an overall coherence of subject coverage, terminology or structure. The resulting mass of texts has come to be called the *Corpus agrimensorum*: the surveyors' compendium. It was a remarkable collection, which not only ranged across surveying and measuring but also incorporated discussions of the Roman water supply and the layout of military camps, maps, itineraries, lists of estates and the history of colonies, agriculture and even a supposed prophecy by an Etruscan nymph. Its contents included geometrical definitions which largely coincided with those in book 1 of the *Elements*. It gave methods for measuring areas such as rectangles, trapezia, triangles, circles, semicircles and hexagons, as well as tips for measuring irregular areas. And it contained classic tricks for using surveying instruments to measure inaccessible distances, such as the width of a river that could not be crossed.

The oldest surviving text of the *Corpus agrimensorum*, written around AD 500, is found today in the Herzog August Library at Wolfenbüttel in Lower Saxony. It is a striking book for several reasons: most obviously because of its illustrations. Some are simple geometrical diagrams in a brownish-red ink. Some are crude pictures to illustrate concepts like the square and the circle. Others look more like the diagrams in manuscripts of Euclid's *Elements*. Others again are more complex coloured pictures, depicting roads (red or brown), hills (mauve), woodland (green) and buildings (brown or yellow, with red roofs). They illustrate the techniques and the results of surveying as it was practised in the later Roman Empire, the geometrical shapes on which it was based and the maps in which it resulted.

This earliest manuscript stands at the head of a long line of later copies made in the monasteries of Merovingian and Carolingian Europe. Its technical Latin sometimes defeated later copyists, and their mistakes and attempts at correction made an obscure text

A geometer in the
Corpus agrimensorum.

more obscure. The diagrams deteriorated too, as scribes did not always check how they were meant to relate to the accompanying text. The actual practice of Roman surveying was fading further and further from memory, as the open-field system of the Middle Ages had no use for the regular square divisions of Roman centuriation. The *groma* and its actual use were becoming matters of merely historical interest.

One of the surprises in the story of Euclid's *Elements* is that the Latin surveyors' compendium came into quite a close relationship with its theoretical cousin. No complete Latin version of the *Elements* has survived from the Roman period, nor does one seem to have gained any real currency or left its mark on the Latin-speaking world. What was eventually produced, perhaps in the fifth or sixth century, was a proof-free summary of the first four books of the *Elements*; it came to be attached – perhaps rightly – to the name of Boethius. By the time the surviving manuscript evidence for it begins, in the eighth century, that summary was invariably circulating in association with the *Corpus agrimensorum*, in a range of different, and intriguing, hybrid texts. In one case Euclid's definitions and postulates, and his first three theorems, were copied into the surveyors' compendium. In another, a Benedictine monk concocted an 'Art of geometry and arithmetic' in five books, by putting together material from Euclid, the surveyors, and other sources on arithmetic. In yet another, a scribe in Lorraine produced a two-book work marrying geometry, the use of the abacus, and surveying.

What was happening was that, in the early medieval schools and monasteries, a whole new discipline was being constructed. It was called 'geometry' and its basic texts were sometimes attributed to Euclid, sometimes to Boethius. Its content was Euclidean definitions, axioms, and statements – but no proofs – together with illustrations, methods and calculations taken from the *Corpus agrimensorum*. It began with descriptions of the plane figures and worked through more complex shapes, together with the ways of drawing their boundaries and measuring their areas.

This new model of 'geometry' was not really meant for practical use in the fields. Rather, in a Christian context, its main use seems to have been as a preparation for the study of theology, just as Plato and Proclus had made Euclid's geometry a preparation for philosophy. Latin scribes and teachers had picked up the connotations of stability and certainty that geometry possessed for many authors, and the specifically philosophical status that it enjoyed for the Platonists. One anonymous Latin author wrote that 'Euclid's intention is twofold: aimed at the pupil and the nature of things . . . At the nature of things, since it is known that the science of nature and the splendid learning of Timaeus or Plato demonstrate geometrically.' Studying these geometrical things – even studying the shapes and measurements of Roman fields – was therefore a way to enlarge the scope of spiritual meditation.

Furthermore, the Old Testament several times spoke of God as a geometer or surveyor: setting a compass on the face of the deeps; stretching a measuring line across the earth's foundation; creating all things in number, weight and measure. If creation itself was a geometrical act, all the more reason to study geometry as a way of comprehending the Creator's mind. Modifying a line from Cassiodorus, an unknown medieval author urged the student 'to approach the heavens with the mind and investigate the entire construction of the sky, and in some measure deduce and recognise by sublime mental contemplation the Creator of the world who has concealed so many beautiful secrets'. The orderly, neatly delimited field of the surveyor became an image of an orderly cosmos; the science of geometry an image of the act of creation.

So the vagaries of translation and transmission had once again brought Euclid's *Elements* to a new location: this time the medieval Latin schools. Here as elsewhere, local scholars and practitioners took the text and made of it something new, sharply extending the range of meanings others thought the text possessed. The boundary between surveying and geometry had been broken down, and Euclid was now the hero of both.

Muhammad Abu al-Wafa al-Buzjani

Dividing the square

Baghdad, in the second half of the fourth century AH (second half of the tenth century AD). A meeting of geometers and artisans; one of a series of such meetings. One of those present is the mathematician Muhammad Abu al-Wafa al-Buzjani.

The agenda is to discuss geometrical problems of interest to the artisans, such as the division of a square into two, three, five or more smaller squares. Such things are of interest for the decorative arts, for making mosaics and designs on walls, ceilings or doors.

The artisans, or one of them, ask how to turn three small squares – of equal size – into one larger square. The problem is perplexing, and a long discussion follows, to which artisans and geometers both contribute their distinctive ways of approaching it.

The glory days of Abbasid Baghdad were in the past by al-Buzjani's lifetime – the old Round City lay in ruins – but the city and the region were enjoying a renaissance of sorts under the Buyid dynasty. The Islamic empire that had once stretched continuously from the Guadalquivir to the Ganges had broken up from the later ninth century, forming a Muslim commonwealth of smaller states, linked by religion and language but separately governed.

The Buyids hailed from Daylam, a mountainous region at the south-western end of the Caspian Sea with a history of supplying mercenaries to the Islamic world's internal struggles. Three brothers, Ali, Hasan and Ahmad, seized power in Baghdad in 945; Ahmad took the position of amir (chief commander) and de facto protector of the caliphs, whose power they effectively usurped in fact. The dynasty ruled Baghdad and its lands – Iraq, plus the family's base in Iran – for a century.

It was a fruitful century for Baghdad's cultural life, though an increasingly calamitous one for its people. Against a background of social and economic decline – particularly in the crowded capital – the Buyid rulers styled themselves 'kings of kings' and pursued elaborate building projects including a new palace, as well as patronising literature, philosophy and science.

The city remained a religious, ethnic and cultural meeting place, and it continued to attract scholars from all over the Muslim world. The Buyids and their viziers supported scholars and poets who consciously attempted to retrieve, transmit and extend ancient culture, translating, editing, commenting and innovating. Ibn Sina (Avicenna) was a product of the period; so was the *Fihrist*, the celebrated Arabic index-cum-encyclopaedia of writers and litera-ture.

It was a culture of informal 'schools': circles of scholars and pupils meeting in shops, city squares, gardens or private homes to discuss and debate. High-profile debates took place in the courts of amirs and viziers, while for theological questions the mosque was the natural place.

One of those attracted to Baghdad and its cultural riches was al-Buzjani. Muhammad ibn Muhammad ibn Yahya ibn Isma'il ibn al-'Abbas Abu al-Wafa' al-Buzjani (generally known as either al-Wafa or al-Buzjani) was born in Buzjan (modern Buzghan) in Iran, in June 940. His family was educated, and two uncles were able to teach him arithmetic and geometry. In 959 he moved to Baghdad, where he enjoyed court patronage and taught and wrote until his death in (probably) 998.

His abilities were much admired by his contemporaries, and to this day al-Buzjani is reckoned one of the greatest mathematicians born in Persia. He ranged across the whole of mathematics: astronomy and the trigonometry that supported it; geometry in the Euclidean manner; arithmetic and algebra. Astronomy was a particular specialism. Al-Buzjani was involved in constructing a new observatory in Baghdad under the amir Sharaf al-Dawla, marking an effective revival of the astronomical tradition in the city. In the later 970s he made observations that would be used for centuries.

He worked on spherical trigonometry: the rules governing the sizes and shapes of triangles drawn on a sphere, crucial for doing astronomical calculations efficiently and accurately. He improved trigonometric tables and showed how to use trigonometry to tell the time from the altitude of the sun. He helped his colleague al-Biruni to determine the size of the earth; the two men coordinated their observations of a lunar eclipse in 997 from Baghdad and Khwarazm.

Twenty-odd publications are mentioned, but only twelve survive. Al-Buzjani's sources were both the Greek classics and the rich Arabic tradition of commentary, development and novelty. He wrote commentaries on Euclid, on Diophantus and on the algebra of al-Khwarizmi: all are lost.

What survives includes certain of his writings on astronomy and trigonometry, and a pair of works written for the use of practitioners. His book on the arithmetic needed by scribes and revenue officials ranged across proportion, multiplication and division, measurement, and the technicalities of taxation, exchange and business transactions. It is said to be the earliest place where negative numbers (for debts) appear in Arabic. Al-Buzjani collected and systematised the methods used by practitioners, and one of his goals was to correct those that needed correction. He noted, for instance, that surveyors routinely found the areas of quadrilaterals by summing opposite sides, multiplying the two sums, and dividing by four: the same traditional procedure that had been applied to the land of Petechonsis a millennium earlier. He wrote 'This is also an obvious mistake and clearly incorrect and rarely corresponds to the truth.'

Al-Buzjani brought the same combination of technical expertise, familiarity with the classics, and willingness to improve and innovate, to his dealings with those who used geometry for practical ends. Mathematical methods were ever more assimilated into the practices of the Islamic world, and Baghdad's culture of meetings and discussions included circles in which geometers like him participated together with carpenters, inlayers and mosaicists; perhaps with surveyors too. As a respected, even revered geometer and astronomer Al-Buzjani enjoyed some authority over both the other geometers and the artisans. The number of these meetings is not mentioned in al-Buzjani's writings, but evidence for them continues sporadically into the next century and appears again up to the seventeenth: it is normally assumed they were a regular occurrence, a reflection of a remarkable, long-lived symbiosis between theory and practice in the Islamic world.

Al-Buzjani wrote up what he had taught and learned from these meetings in his 'Book on what the artisan needs of geometric constructions' in the 990s, dedicating it to the amir Baha' al-Dawla. Like his arithmetic book, it was meant not only as an introductory textbook but also as a corrective to some of the errors commonly made by professionals:

> A number of geometers and artisans have erred . . . The artisan and the surveyor take the choice parts [literally, the cream] of the thing, and they do not think about the methods by which correctness is established. Thus occur the errors and mistakes . . . The geometer knows the correctness of what he wants by means of proofs, since he is the one who has derived the notions on which the artisan and the surveyor base their work. However, it is difficult for him to transform what he has proved into a [practical] construction, since he has no experience with the practical work of the artisan and the surveyor.

Al-Buzjani was able to move in both worlds and could correct sloppy artisans and unworldly geometers, to the advantage and use of either. His work and his career thus became a crucial moment in the application of Euclidean geometry to real-world problems, particularly those of design.

Much in his book derived from Euclid, Archimedes, Heron, Theodosius and Pappus: the heroes of Greek geometry known in Arabic translations and commentaries. Much, though, was original, and al-Buzjani was fond of extending Euclidean methods by the use of what is sometimes called a 'rusty compass': a compass that would keep its arms in a fixed position between uses (as the traditional compass of Euclidean geometry would not), allowing lengths to be copied from place to place in a diagram. This subtle change allowed him some new results and greater precision in the solution of practical problems.

Much in the book derived from the practitioners themselves; some of the problems were familiar ones from carpentry: testing that a given angle was a true right angle, for instance. Al-Buzjani wrote about the design and use of geometrical instruments, and the construction and measurement of a wide range of figures, shapes and designs. How to divide a line into equal parts, for example; how to find a line that just touched a given circle, or draw regular polygons inside or outside a circle; how to draw polygons inside or outside each other – a triangle in a square, or a pentagon in a triangle; how to divide the areas of given shapes in various proportions. Two particular interests late in the book were how to divide a given square into a certain number of other squares, and how to divide the surface of a sphere using regular polygons.

Tessellation, whether of planes or spheres, is one of the features most associated with the tradition of design that was sweeping the Islamic world during the later tenth and eleventh centuries. Towers were decorated with intersecting polygons in cut brick, like those at Kharraqan and Maragha in Iran; doors or chests were decorated with similar designs in wood. These designs filled spaces harmoniously without cluttering them; they generated large-scale designs from simple elements applied over and over. Both the processes and their result have been called symbolic of the characteristics of the cosmos: harmonious, unified and relational.

There was interest in such things in ancient Greece, too. The Greek world had a game called the *stomachion*, that consisted of a set of flat ivory polygons to be rearranged into various shapes; Archimedes wrote about it, and there were Arabic versions of at least part of his text. There was a family relationship with some of the proofs in book 1 of the *Elements*, which dissected the figures they dealt with, and in fact Euclid himself wrote a whole book, now lost, called *On the Divisions of Figures*. That book was known in Arabic, and chapters 8 and 9 of al-Buzjani's text have every appearance of being written with this Euclidean *Divisions* before him.

The division of squares was the most extensive single topic in this part of al-Buzjani's book. For mosaicists, both the piece to be cut and the finished pattern seem frequently to have been square, and al-Buzjani went through a series of different versions of the problem, forming large squares out of various numbers of smaller squares. Sometimes it was enough to cut the small squares diagonally and rearrange the pieces; sometimes much more complex dissections were involved, typically leaving one square whole and cutting up the others, arranging their pieces in a symmetrical pattern around the central whole square.

Al-Buzjani gave general principles as well as detailed solutions such as for five, eight, ten or thirteen small squares. It was in this context that someone among the artisans asked for a way to make *three* squares into one larger one. Evidently recounting an actual discussion in his book, al-Buzjani recalled that a geometer could easily find the *size* of the larger square, but the artisans were not satisfied with that piece of theoretical knowledge. They wanted an actual cut-and-rearrange solution that they could use in mosaic work.

The artisans, indeed, had a number of proposals, but the geometers were not satisfied with them, thinking them inexact. Imprecise solutions created what looked superficially like a whole square, but in fact had jagged edges or pieces sticking out at the corners. Al-Buzjani demolished two such solutions, before going on to report a better one, most likely of his own invention. The basic idea was familiar: leave one of the small squares whole and cut the others into triangles, which could then be arranged symmetrically around the edges of the whole one. What resulted had both gaps and protrusions compared with the true large square that was wanted: but the protrusions and gaps were the same size and shape, meaning that al-Buzjani could cut off the one and use them to fill the other. The result was a perfect square made from the pieces of three smaller squares: a strikingly elegant piece of mathematical work, and a beautiful design of real value in decoration. It is still in use among stonemasons from Anatolia to Morocco.

Dividing a square.

Al-Buzjani's book was widely circulated, and it was translated into Persian twice from its original Arabic. There were later writings by other authors which took the division and interlocking of figures much further. On the other hand, mathematicians who taught geometry to artisans were still complaining about their ignorance six centuries later.

The impact of the books and the meetings on artisans is hard to measure, because his book is almost the only specific evidence of artisans' exposure to theoretical geometry in this period. But there can be little real doubt that the encounter of Euclid's *Elements* with Persian and Islamic design was a fateful one. Some of the patterns al-Buzjani described offered little potential for producing aesthetic results, but his five-squares pattern and its extension into a nine-squares version were widely taken up, and can be seen on walls, portals, minarets, doors and chests from across the medieval Islamic world: the earliest surviving example is on the early twelfth-century wooden door of the mosque of Imam Ibrahim in Mosul. Six centuries later it was still in use, built into the tiling pattern in the Friday Mosque in Isfahan.

Lady Geometria

Depicting the liberal arts

Chartres, France, around AD 1150. Among the famous stonework figures on the cathedral's west facade, a woman sits writing, her tablet on her knees. A robe drapes her, and more drapery covers her hair. Her sharp features, it seems, gaze outwards; not at her work but at something in the middle distance, outside the frame and outside the cathedral. She neither smiles nor frowns.

Her name is Geometria, and beneath her feet sits another stone figure: one of her votaries, it would seem. Tradition calls him Euclid, probably rightly.

Euclid (if it is he) is also robed, but his hair and beard are free. He has a lap desk to work at, and both his hands rest on it. An inkwell hints at the possibility of writing something, but like Geometria he seems to gaze outwards, matching her line of sight and her pensive, distant expression.

Today, after the better part of 1,000 years, Euclid and Geometria are now weathered and hard to read. Her right arm is gone, and

Geometria and Euclid at Chartres Cathedral.

with it possibly a writing or drawing instrument. The features of both figures are a little blurred. Nevertheless, the sculptures at Chartres are, it seems, the earliest depiction of Euclid to have survived, and the first sculptural version of the seven liberal arts, of whom Geometria is one.

Chartres Cathedral rises, an almost unbelievable sight, from the Beauce Plain between the Seine and the Loire: an artefact as if from another world, suspended between heaven and earth. Visible for a day's walk in any direction, it was a place of pilgrimage: one of the most popular in Western Christendom. The Romanesque church of 1020 was destroyed by a pair of fires in 1134 and 1194; the new, current building was the result of about

| 203 |

a century of reconstruction and decoration from the 1140s to the 1230s. It was an immensely influential high point of Gothic art and architecture.

The west facade at Chartres was sculpted with three elaborately decorated doorways, depicting the Incarnation, Ascension and Second Coming of Christ, garlanded with an unprecedented collection of biblical and secular scenes and figures. The southern portal, on the right, showed the infant Christ in the lap of Mary, who was enthroned in a style sometimes called the Seat of Wisdom. Angels formed a first surround, and beyond them were two ranks of the personified liberal arts and their exponents. It was the first time the personifications and practitioners of secular learning had been used to frame a theological scene, and it identified Christ and/ or his mother as Wisdom, whose children the liberal arts traditionally were.

The sequence of the arts started at bottom right with Grammar, who was teaching – or trying to teach – two boys. One had flung off his clothes and was pulling the other's hair, and Grammar herself wielded a rod or possibly a whip. Not perhaps the best start, but the remaining six arts appeared to be doing rather better. Logic held the symbols of good and evil: a flower and a monster. Rhetoric spoke, Geometry wrote or drew, and the now damaged Arithmetic most likely counted, perhaps on her fingers. Astronomy looked at the sky and Music played a set of bells.

Each art had a companion: a historical person who wrote about the subject. They were not labelled, and some of the traditional names may be wrong guesses: Donatus, Aristotle, Cicero, Euclid, Pythagoras, Ptolemy, Boethius. They were all shown writing or thinking; doing was left to the arts themselves.

Why the seven liberal arts, and why here? One of the most famous scholars associated with Chartres – Thierry – finished his book on the seven arts under the title *Heptateuchon* around the same time the west facade was sculpted. Thierry's teaching emphasised, in what has been called a kind of prototype of the Renaissance, that secular and even pagan learning was inspired by divine Wisdom, and could be a route for scholars to approach that wisdom. It was his brother Bernard, another author, who first famously said that modern scholars were like dwarves sitting on the shoulders of those giants the ancients. The liberal arts, then, were the means by which one prepared oneself for understanding moral and theological truths. Like secular evangelists, they took inspiration from God and passed it on to human beings.

The scheme of sculptures at Chartres placed Geometry and Arithmetic right at the top of the archway, closest to heaven, and perhaps hinting at the common view that mathematics was specially related to philosophy or theology. But Thierry himself relied for his geometry on a mix of Euclidean and practical material. His *Heptateuchon* drew on Adelard's Latin version of the *Elements* but also, heavily, on the *Corpus agrimensorum*, quoting from five of its authors as well as giving much of its truncated version of Euclid.

△

Lady Geometria herself, and the seven liberal arts, were not quite new as a decorative scheme at the time of Chartres, even if they had never been sculpted before. There were two-dimensional precedents including probably in the palace of Charlemagne, church floor mosaics and even the bedchamber of Adela, Countess of Chartres, around 1100. But they were never common compared with other decorative schemes such as the Virtues and Vices or the personifications of the calendar months. The earliest known picture of Geometria is in a manuscript of Boethius' *Arithmetic*, made around 840 at Tours for Charles the Bald. There, Geometria holds a stick in her right hand and stands by a raised surface on which geometrical figures have been drawn: possibly a writing board but perhaps more probably a sand tray, the 'abacus of dust' mentioned by Martianus Capella in his description of the arts and their accoutrements.

Geometry teaching.

A trickle of other depictions followed: a Swabian manuscript of Isidore of Seville; stained glass at Laon; treatises on the arts and learning through the thirteenth and fourteenth centuries. In a fourteenth-century copy of Adelard's version of the *Elements* some of the initials were illuminated, and the P at the start of the first definition contained not just Geometria – once again drawing shapes on a raised table – but a small crowd of her eager pupils. The fresco of the *Triumph of Thomas Aquinas* in Santa Maria Novella in Florence included the liberal arts; so, in the next century, did the stained glass made for the chapter library at Chartres itself.

The personified arts were usually – not always – women, partly for the simple reason that 'art' is grammatically feminine in Latin. Their attributes – the pieces of equipment with which they were pictured – were slow to settle into a fixed pattern. For Geometria there was often a rod, presumably meant for measuring and sometimes definitely recalling the rod with which St John measured the temple and the angel measured the new Jerusalem in the Book of Revelation: sometimes dignified as a special 'geometrical rod', it typically looks like a mere stick. Occasionally she held up a diagram, almost invariably the one from the first proposition in the *Elements*; more frequently she appeared to be drawing simple figures such as circles and triangles for the benefit of the spectator or – as in the illuminated Boethius – her students. By far her most frequent attribute was a pair of compasses, and eventually a fairly settled image of Lady Geometria was arrived at: instead of holding them up or waving them vaguely about, she leaned forward to trace a circle on the ground. For this to work, either the compasses had to be huge or Geometria had to bend almost double, but either way the compasses were shown in actual use and the ancient image of the geometer tracing figures in the dust was regained. (It remains a favoured way to depict geometry in action: the monumental sculpture of Isaac Newton outside the British Library in London, built in 1995, adopts the same pose.)

The accoutrements of the liberal arts fairly often included not just objects but people: exponents of the arts, or practitioners of them. No fixed cycle of personnel ever emerged for this purpose, and some arts varied among several companions: but the geometer was almost invariably Euclid. An alternative to exponents was to give the arts assistants – practitioners – transferring the practical doing of the art away from its personification and into the human realm;

in the case of geometry that resulted in Lady Geometria being flanked by carpenters or surveyors. In some versions the art and the practitioner were conflated: or the exponent and the practitioner. (A relatively rare tradition of depicting a cycle of 'practical arts' in addition to the liberal arts sometimes confused matters still further, with carpentry or surveying receiving personifications and attributes of their own.)

The seven liberal arts.

A particularly striking depiction of the liberal arts appears in a fifteenth-century manuscript of the *Clock of Wisdom*, in a picture showing both an idealised school of theology and the liberal arts in a space of their own. The seven ladies were here provided with a sort of library, with several books on a central stand. They stood or sat reading or working; they calculated, observed and wielded an astrolabe. Grammar taught four pint-sized men who huddled at her feet. The stars were visible through an archway. Geometria applied her usual stick and compass to the floor, but she also had beside her a specimen of that very characteristic Roman surveying instrument, the *groma*.

And so it went with Euclid himself; over time the depictions of him veered between the cloudy sage and the earthy practitioner. One fifteenth-century German manuscript made him a mason; another seemed to show him working at a carpentry table, with assistants who wielded a stick and a plumb line. The caption said that 'by measuring and weighing by means of triangles and the like, geometry is here being carried out'.

Another early depiction of Euclid made him a different sort of practitioner: an astronomer. One manuscript now in the Bodleian Library in Oxford was copied in probably the 1230s or 40s by the celebrated scribe and scholar Matthew Paris in the monastery at St Albans. It is a parchment codex of seventy-odd leaves, and it contains a miscellany of texts on prognostication and divination – fortune-telling – of the kind called *sortes*, which employed random processes to answer questions for an enquirer. Typically there was a table of available questions and a table of their possible answers; the client chose a question and the master rolled the dice, drew the lots or stuck a pin in a book in order to start a random process which would choose one of the answers.

In the well-known cousin of such processes called the *sortes Biblicae* – choosing a Bible passage at random – it was implied that divine providence determined the choice. In more secular situations it could be implied that other numinous powers including the stars or planets were at work, and so certain *sortes* were characterised as astrological, even though they need involve almost no knowledge of the movements of the heavens.

The various *sortes* in the Bodleian manuscript are introduced by portraits of their (alleged) authors with suitable attributes. There is a Socrates and a Plato, there is a Pythagoras and a half-page with the twelve biblical patriarchs. Other decorations include birds and animals, flowers and fruits. But the whole volume begins with a half-page picture of two seated men labelled 'Euclid' and 'Hermann'.

Euclid and Hermann.

Hermann was an alleged collaborator on one of the texts in the volume: he was Hermann the Lame from the eleventh century, remembered today as an author of hymns but better known in the Middle Ages, among other things, for his treatise on the astrolabe. (It was a small world: Hermann also once addressed a work to Thierry of Chartres.) But for Euclid there is no comparable connection with the accompanying text: he is neither mentioned in the *sortes* nor ever said to have been their author. One reason for depicting him was that the start of the first text vaguely recalls the start of the *Elements*: 'first let there be a line, at random, made of a number of points . . .' Another was the generally astrological or astronomical cast of some of the material; since Euclid had in fact written on astronomy, the scribe possibly felt that he was as good a candidate as any to represent the distinguished history of astronomical writers.

Whatever the exact reasons, there he was placed. Hermann is shown offering him an astrolabe, appropriately enough, while Euclid himself does what astronomers are naturally depicted doing: he looks at the stars. He sights along what might be either a rod or a tube; some readers have seen it as a *dioptra*, the Greek surveying and sighting instrument functionally equivalent to the Roman *groma*, but it is not clear it is meant to be more than a simple rod. In his other hand, his right, he holds a round object probably meant as a spherical model of the heavens: a small armillary sphere. His curly hair and beard and his Phrygian bonnet mark him out as vaguely 'oriental' (the characteristics were most often associated in medieval art with depictions of those other wise men from the East, the Magi of Matthew's Gospel).

Euclid, and his image, were becoming common property, general-purpose symbols of mathematical learning and endeavour fit for a wide variety of contexts. Robed, gowned, dressed up and given tools and instruments to wield, he frequently seems to have been something of a hero for geometers and those who thought about them: revered, but on the whole perhaps, misunderstood.

Piero della Francesca

Seeing in perspective

A bearded man stands by a pillar, his back bare to the whips of two soldiers. A man in costly clothes watches the scene from his seat to the side; another observer stands. Outside the building, closer to the viewers, three more men converse. They seem oblivious to what is taking place behind them, though the way they are grouped seems to echo the soldiers and their victim, and the fantastic variety of their clothes echoes a similar variety in the main scene.

Despite the violence of the events, the scene has an unearthly calm to it. The light defines the figures, but it is hard to read: there is perhaps a lamp shining from somewhere the viewer cannot see. Order and chaos seem balanced in a way that is hard to grasp; and the two groups of figures must, surely, have something to do with one another: but what?

The *Flagellation of Christ*, painted by Piero della Francesca in – probably – the 1450s or 60s, was a world away from the stylised

Piero della Francesca, the *Flagellation of Christ*.

depictions of geometry and of Euclid from the Middle Ages. But it possessed geometrical qualities that linked it closely to Euclid and his writings.

Following from Euclid's *Optics*, there was a rich tradition of writing about optical matters in later Greek, Arabic and Latin, and by the turn of the fifteenth century horizon lines and vanishing points were being used by artists to provide some illusion of a viewer's position in relation to the depicted space: up or down, left or right. The Latin word *perspectiva* so far simply meant the science of optics, but it would be transferred during the Renaissance to the science of illusionistic drawing. Filippo Brunelleschi, the artist and architect who designed the dome of Florence's cathedral, is credited with inventing a rule that helped locate the eye in the third dimension, though it is not at all clear what his rule actually was.

Leon Battista Alberti, another architect, included a few pages about perspective in his treatise on painting – dedicated to Brunelleschi – in the 1430s: but it was a wide-ranging book meant for connoisseurs to read, not for artists to learn from, and on the technical side its discussion was both unclear and, by Alberti's own admission, incomplete. His general disposition, like that of many of his Italian contemporaries, was that geometry and its study of proportion could put the description of beauty on a rational foundation; that geometrical harmony was the answer to most of the worthwhile questions about the arts. But how exactly were artists to get from the pure forms of geometry to the messy shapes of actual experience? Alberti left the question largely open.

Enter Piero della Francesca. His *Flagellation* is one of the classics of early perspective, but it is enigmatic in most of the ways a picture can be. It has been dated as early as 1445 and as late as 1475, making it an early work of della Francesca's or a late work or anything in between. As to size: at very roughly two feet by two and a half, it is small for an altarpiece or an independent panel, but too large for a predella. The detail with which it is painted demands to be viewed from up close – you can imagine a connoisseur literally holding it close for inspection – yet the perspective is constructed in such a way that it only looks correct from more than six feet away.

The design is no less mysterious, quite different from other contemporary representations of the flagellation of Christ. The left part of the picture has a fairly traditional group, with Christ at the pillar and two soldiers plying their whips. The seated, watching figure is naturally read as Pontius Pilate; the other watcher is hard to identify. But the matching (does it match?) group on the right seems quite a separate scene, perhaps from a different place or time. The painting has been called a story of two voices on one picture, or two narratives, pulling the viewer's gaze in different ways. The two scenes appear to be at drastically different distances from the eye, making the right, foreground figures dominate by their sheer size on the panel and giving the actual flagellation the air of something seen distantly, through a window.

Finally the lighting, too, is a puzzle. Della Francesca was immensely skilled at handling the fall of light and the play of colours, and this is a beautiful picture. But the two scenes are lit differently. The foreground has sunlight from the upper left; the background seems to have a separate bright source of light between two of the columns, perhaps just where Christ is looking.

Attempts to make sense of all of this have ranged widely. There are relationships between the sizes of objects in the picture, such as the squares on the floor and ceiling; there are also relationships between their distances from the eye: for instance, the watching figure in the turban appears to be almost exactly twice as far away from the viewer as the three figures in the foreground. How much of this was deliberately constructed by the artist is difficult to guess. The square pattern of the floor on which Christ stands is cut by a circle, which might just be a gesture towards mathematical complexity, irrationality, absent from the simpler floor in the foreground. But it also has a possible relationship with pilgrims' accounts of the Holy Land: the four columns, the marble floor and its central disc can all be found in their descriptions of the locations of the Passion.

Meanwhile, the three foreground figures have been reinterpreted almost as often as they have been written about (a recent article lists twenty-nine published identifications of the figures, of which no two agree). Their mix of Greek, Italian and classical (or faux-classical) clothing has invited both historical and symbolic readings: the pope, the Byzantine emperor, the Duke of Urbino, a youth miraculously raised from the dead by St Helena; or generic

– theologians, astrologers, peasants, merchants . . . If the painter had a specific intention, it seems impossible to recover it with certainty.

With so much mystery in a single panel, it is a relief to turn to the painting's perspective. As one historian puts it, this is not *an* example of correct perspective from this period, but *the* example. For della Francesca's generation, perspective was an established method, and it is clear that his understanding of it went well beyond anything he could have learned from Alberti's book. Lines meant to be read as perpendicular to the picture plane all converge to a single point on the back of the judgement hall. The ground plan and elevations of the scene can be reconstructed in detail from the picture; and della Francesca surely drew a plan and an elevation for reference before he painted the scene. The detailed pattern on the floor, too, with squares divided into squares, must surely have been designed in plan before it was rendered in perspective.

The sense of stillness that many writers have found in della Francesca's paintings, and in this one in particular – a stillness that sharpens by contrast the violence of what is depicted – has, perhaps, something to do with this attention to the perspective. From the ideal viewpoint, the viewer really can take in the whole scene at once, with no need to fumble incorrect elements into place or to move around to deal with multiple vanishing points. It is exceedingly rare – perhaps even unique – in fifteenth-century art for such a correctness to be achieved.

For all that, the picture is not a geometrical diagram. The small squares vary slightly in width; the ceiling beams are not exactly parallel and the uprights not quite vertical. The path that comes out to meet the picture plane is narrower than the other paths in the picture. This is a real house, not an architect's idealisation.

Piero della Francesca came from a small town in the upper valley of the Tiber. He was painting in his native Sansepolcro by 1430, in Florence by 1439. Painters in the Italy of his day were craftsmen – artisans – and were educated accordingly, learning basic mathematics and (vernacular) reading and writing in local schools. Della Francesca most evidently had an aptitude for mathematics, as well as an exceptional skill at visualising and composing in two and three dimensions. During his life he acquired a deep knowledge of both Euclid's *Elements* and Euclid's *Optics*; the interest was usual, even traditional for an artist, but the depth with which he pursued it was perhaps not.

Neither was the fact that he wrote on mathematical subjects. Three of his mathematical treatises have survived. The first, probably from fairly early in his life (though the proposed dates range over a thirty-year period), was an 'abacus book', the *Trattato d'Abaco*. It was the kind of textbook that would have been used in local schools, and it covered the subjects of arithmetic, algebra (done in words) and geometry. It consisted of long series of problems of a 'practical' kind: some really practical, most rather contrived so that the numbers came out well or a mathematical point was made. There were inheritances to be divided and tanks to be filled with water, ladders to be leaned against walls and fields to be measured: all elements in a dense tradition which in Italy went back to the 1202 *Liber abaci* of Fibonacci, which in turn drew on Arabic models.

Della Francesca's *d'Abaco* contained more geometry than most, and it quickly left behind practical problems of measurement in favour of rather abstract questions about polygons and polyhedra, which were evidently what interested him best. For instance: there is a triangle, whose sides are thirteen, fourteen and fifteen units, and you want to draw inside it the largest circle possible. How large is that circle?

The book worked through questions about squares, rectangles, pentagons, hexagons and octagons, and even discussed the regular solids, uniquely among abacus books of its time (and possibly with a deliberate nod to the culmination of Euclid's *Elements* in the same solids). You have a sphere whose diameter is seven units and you want to put inside it a figure whose faces are four equilateral triangles, so that its corners touch the sphere. How long will its edges be?

It was like the *Elements* but also unlike it. Della Francesca, indeed, was not just a reworker of traditional ideas but a mathematical innovator. There were problems throughout the *d'Abaco* that seem to have been originated by him, and when it came to the three-dimensional solids he introduced two intriguing figures which he had apparently (re)invented for himself: one with eight faces (four triangles, four hexagons) and one with fourteen (six squares and eight triangles).

Della Francesca, then, had a deserved reputation for mathematical skill, and it is no real surprise to find that he also put it to use in writing an account of the theory of perspective: the first substantial such account, in fact. Once again, there is uncertainty about when he did so, but a date around 1460 seems most likely. His book was translated into Latin perhaps twenty years later for presentation to the ducal library in Urbino. Unfortunately the Italian original has been lost.

The book – *On Painting in Perspective* – may have been prompted by requests for explanation from della Francesca's contemporaries, but it is doubtful how many among painters would have actually understood it. Compared with Alberti's general treatise on painting it was a very different type of book, focussed specifically on perspective in its technical aspects. The first section gave theoretical preparation and dealt with the perspective of points, lines and surfaces. The remainder of the book provided practical ways to make perspective pictures. By and large the style was that of the *d'Abaco*, with detailed worked examples set out as step-by-step instructions for the reader, who was addressed familiarly as '*tu*' and obviously expected to follow along, with drawing implements in hand. The reader's hand was closely guided through the steps needed to construct the supporting diagrams, in a manner surely derived from the oral instructions given to apprentices in a painter's workshop: effectively reducing the learning of perspective to the copying of a given set of images. Yet the book fused this practical manner with that of Euclid.

The book was set out in a series of 'propositions', indeed, and several of the early ones were taken directly from the Euclidean *Optics*. If two objects are of equal size, for instance, the one nearer the eye will appear the larger. Della Francesca went beyond Euclid by systematically thinking in terms of a 'screen' placed between the eye and the viewed object, onto which the image was supposed to be projected. In effect the question became not 'what angles will these objects make at the eye?', but 'what, drawn on this screen, will look the same, to the eye, as the real objects?' The relationships between shapes and sizes on this screen were geometrical, and could be investigated and specified using Euclidean methods. Della Francesca defined painting as a branch of geometry, and perspective as a science of measurement: a way of seeing that was superior to mere natural vision, recalling Alberti's enthusiasm for geometrical harmony and proportion.

So della Francesca placed himself at an intersection of two traditions, linking the worlds of craftsman and scholar. Unlike al-Buzjani he was not a scholar in dialogue with artisans; both functions were here merged in the person of the painter himself. From him you could learn not only the proportions in which distant objects grew smaller to the eye, but also how to draw a house, a column, a well in correct perspective. He showed not only how to

shrink a square in a given ratio or how to place a correctly perspec-
tived cube on top of it, but also how to retain the proportions of
churches, temples, round windows or even human heads, when
they moved and lay at various orientations in a picture.

Della Francesca's own pictures were certainly a source for his
Perspective. The key diagram in part 1 of the book showed a large
square containing several smaller squares receding to a vanishing
point, with a schematic observer viewing the scene from the side.
It was virtually a mirror image of certain elements from the
Flagellation. The square patterns on the *Flagellation*'s floor and
ceiling also appeared, slightly altered, in the *Perspective*, and so
did the stripes of white marble in the flooring. Even the plinths,
the columns and capitals from the painting had close relatives in
the book. There were borrowings from della Francesca's other
paintings too, underscoring – for those who knew where to look
– the unity of his art and his mathematics.

Della Francesca was not yet finished with mathematical writing.
Towards the end of his life, probably in the 1480s, he composed
another treatise, this time on the five regular solids. Dedicated to
the new Duke of Urbino, and in due course translated into Latin,
it reveals that della Francesca continued to immerse himself in
Euclidean material after the *Perspective* was finished.

The book – *On the Five Regular Solids* – copied or reworked
material from the *d'Abaco*, and went on to deal with the solids in
still more detail, effectively reproducing the content of *Elements*
books 13 and 15. He discussed how to place one polyhedron inside
another, and what relationship of side lengths would result. As with
the *d'Abaco* and the *Perspective*, it was all done in worked examples,
with specific numbers put in for the sizes of the polyhedra in place
of Euclid's abstract lengths. Find the diameter of a sphere that

contains a regular tetrahedron whose side measures four bracchia. Find the base of an icosahedron whose volume is 400.

Like the *d'Abaco*, too, this book introduced some irregular solids which seem to have been his own (re)discoveries. Della Francesca wrote about five, including four new ones compared with the *d'Abaco*: the most spectacular had seventy-two faces: twenty-four triangular and forty-eight square. His method of proceeding was to start with a regular solid and cut off its corners in such a way as to turn, say, triangular faces into hexagons, or square faces into octagons. If it was done in the right way, the resulting new faces would be regular polygons, and thus the solid 'semi-regular', with the same set of polygonal faces arranged around each of its corners. This method of 'truncation' seems to have been another of della Francesca's inventions.

The *Regular Solids* ended with a series of afterthoughts, including the strikingly practical problem of finding the volume of an irregular object by immersing it in a measuring cylinder: a good Archimedean trick, and one that quietly emphasised that this was not the purely Euclidean treatise the title – and the dense network of Euclidean citations inside the book – might suggest. Like the *Perspective*, in one sense it was Euclidean but in another not; della Francesca seems to have been determined to carry on Euclid's legacy in his own particular way.

Piero della Francesca died in 1492, and his books were not published in his lifetime. Nevertheless, his *Regular Solids* seems to have been widely read, and all three books were influential through their reuse (to put it delicately) by mathematician Luca Pacioli in the next generation. Pacioli's long career throughout Italy brought him into contact with Alberti and della Francesca among many others, and he shared the patronage of the Duke of Urbino with the latter. After della Francesca's death he seems to have acquired his manuscripts, and the sixteenth-century writer Vasari was scathing about his publication and use of them in his own books without giving proper credit.

Pacioli's book on arithmetic made use of problems from the *d'Abaco* as well as its treatment of the solids and material from the *Perspective*; Pacioli did credit della Francesca with putting perspective on a geometrical foundation, but that was all. His *Divine Proportion* included an Italian version of the *Regular Solids*, also without attribution. Pacioli did much to make the study of the geometrical solids fashionable, and he was famously painted teaching the Duke of Urbino with two geometric solids – one wooden, one of glass – hanging in the background. The portrait clearly depicted Pacioli as a Euclidean expert, with the *Elements* open at an annotated page of book 13 and a set of drawing instruments offered to the viewer, as though to take up the challenge the book represented.

Another of Pacioli's famous friends was Leonardo da Vinci (they shared a house in Milan), who provided illustrations of the solids for the *Divine Proportion* which amounted to fifty-nine full-page woodcuts when the book was printed. Where della Francesca had given simple linear sketches in his manuscripts, Leonardo provided richly visual, even tactile pictures of the solids, complete with light and shade. He also provided a 'skeleton' version of each, with the edges reduced to thick sticks so that viewers could see through the solids and glimpse their construction more fully. In some cases the illustration gave substantially more information about the solid than did Pacioli's text. Neither della Francesca nor Leonardo, however, got the perspective of the solids perfectly right.

Through Pacioli, the study of the solids was taken up elsewhere, by such figures as Albrecht Dürer and Wentzel Jamnitzer in Germany. Dürer described seven irregular solids, of which five had appeared in Pacioli's works and two were new; and instead of showing illustrations he provided patterns, 'nets', from which the reader could construct a solid model out of paper. It was perhaps the logical next stage in della Francesca's fusion of Euclidean geometry and manual practice.

Jamnitzer's *Perspective of the Regular Bodies*, published at Nuremberg in 1568, was a sort of culmination. It showed perspective illustrations of around 150 different solids, all derived from the five regular ones. He truncated solids; he 'stellated' them by adding a pyramid to each face; he derived solids from solids without any very obvious system: and with little accompanying text it was hard to follow his line of thought. He gave no instructions for drawing or constructing the solids: just a seemingly endless series of pictures. He devoted a chapter to each of the Platonic solids and its derivatives, and each was introduced with an illustration of the element with which that body corresponded in Plato's ancient cosmology.

The purpose of Jamnitzer's book was arguably to look impressive rather than to convey any very definite information. More kindly, it can be seen as a pattern book for artists and artisans. The geometrical solids – and their representation in perspective – had become common property, part of artists' and intellectuals' self-image quite generally. There was a fashion for displaying them – in portraits, marquetry, games boards or sundials – to show that one was both learned and skilled, to display one's ingenuity and one's elevated mind at the same time; there was a vogue for using them as decorative elements on tombs.

A fusion of geometrical, Euclidean learning with practical craft skills had become – thanks in quite large measure to Piero della Francesca – part of the European landscape. The profession of the so-called 'mathematical practitioner', dextrous with both mind and hand, would become increasingly established during the sixteenth and seventeenth centuries. As far as art was concerned, the technique of perspective became a basic part of a young painter's training, a prerequisite for even the least ambitious artist. And – even if Piero's book on the subject had dropped almost out of view – nearly every other treatise on perspective dutifully adopted his series of worked examples to teach how perspective drawing worked and how to do it.

Euclid Speidell

Teaching and learning

In London, during the 1680s, Nathaniel Denew takes lodgings in a family home in Whitechapel. His landlord happens to be a mathematical practitioner, and Denew takes lessons from him for a time. He learns arithmetic: the manipulation of numbers, of weights and measures, and particularly of money, in the complex systems of units in use in England at the time.

The teacher and landlord makes much of his income measuring and estimating the volumes and quantities of goods for the purposes of taxation. But he also dabbles in mathematical teaching, in public houses and rented rooms, including at the King's Head in Threadneedle Street. Later in life he publishes a few mathematical books. His name is Euclid Speidell.

How did Denew – probably a young man, and possibly new to London – feel about learning mathematics from 'Euclid'? He would have addressed his teacher as Mr Speidell, but still. Imagine learning philosophy from a teacher called Plato.

Practical mathematics was embedded in the culture of sixteenth- and seventeenth-century Europe. Mathematical practitioners did everything from drainage to tax surveys to fireworks to fortification to sundials; they taught, wrote, printed, sold books and plied their mathematical skills in every way they could think of that people would pay them for. Many of them adopted Piero della Francesca's habit of using Euclidean ideas and methods, and so they gave Euclid and his *Elements* a new visibility and a new reach, as practical mathematics generated its own enormous body of texts, each one responding to a different combination of local concerns and situations.

If mathematics was an island of order in a chaotic world, what better way than to appeal to the special kind of certainty and truthfulness that some philosophers said attached to Euclidean methods of proof? If your identity as a practical mathematician depended on the authority of mathematics, what better way to build up that authority than to appeal to the grand ancient *Elements*, now available in print in editions to suit every need? And if your income depended on teaching mathematical methods, what better way to look competent and attractive than to adopt the structure of knowledge implied by the systematic arrangement of Euclid's *Elements*?

The people involved were, as one historian puts it, 'a new sort of intellectual', who 'moved between workshop, university and court', and 'built their personae around the mathematical practices such as drawing architectural plans, calibrating horoscopes, surveying, fabricating instruments, taking observations, and teaching students'.

The books they wrote covered all these topics and more, and so Euclid's definitions, his methods and a few of his propositions – or references to them – found their way into a gamut of practical books: manuals of perspective, practical and popular works on astronomy, works on surveying and measurement. Hydraulics, architecture, even fencing, dance and dressmaking found their systematic, 'geometrical' presentations in books in English, French, Dutch, Italian, German and Spanish claiming to be based on Euclid but offering real practical results. They were sold, reprinted, taken to classrooms and workshops and actually used. One *Young Mathematician's Guide* from the eighteenth century has sawdust still wedged between its pages.

Euclid Speidell lived from 1631 to around 1702, and he exemplifies the middle-ranking mathematical practitioner to whom Euclidean ideas now belonged. As well as the occasional testimony of his contemporaries, and the books he published, his manuscript autobiography survives: it runs to sixteen pages and allows quite a close view of his life.

His was a German family; Euclid's grandfather seems to have moved to England in the reign of Elizabeth I and worked at the Royal Mint. His father, John, was a successful mathematical practitioner, publishing various books on subjects such as arithmetic and logarithms; he invented a kind of rod for determining how much liquid a barrel contained. John taught and published on mathematical subjects in the first half of the seventeenth century, offering lessons at his home 'in the fields betweene Princes streete and the Cockpit', as his son Euclid would later do. He taught in French, Latin or Dutch as the need arose, and he advertised his teaching as including arithmetic, algebra, geometry, astronomy, navigation, fortification, the making of sundials, trigonometry and the use of logarithms. For good measure, he also sold students copies of his books and 'the best Mathematicall paper'. During the first English civil war, John Speidell's practice declined, and he died in 1649, leaving six surviving children.

John Speidell was the kind of mathematical practitioner who

often cited Euclid, and who adopted Euclidean or quasi-Euclidean methods in his writing. Two books of mathematical problems he published had a thoroughly Euclidean flavour. Nevertheless, the choice to name his son after the great Greek mathematician was unusual, and it is not absolutely certain whether the young man was given the name at birth or adopted it later as a professional badge. For him, periods of school education alternated with periods of education at home. There was a spell at Westminster School, under the well-known headmaster Richard Busby who had offered the promising young man free tuition; young Euclid recalled mainly the learning of Latin grammar and verse composition there, and his autobiography conveys rather a sense of relief when he was able to return home and learn more arithmetic and surveying from his father. In 1648 they took a survey of the woods at Eltham in Kent.

By the age of thirteen he was teaching other children arithmetic, but his father died when Euclid was eighteen, and his life over the following decade was decidedly peripatetic. He trained and worked as a legal clerk; he went to sea for two years, presumably as a captain's clerk. Having had his fill of shipwrecks and sea battles during the Anglo-Dutch wars, he then returned to land and worked as a steward. He taught mathematics, and eventually made a career as a mathematical practitioner in his own right. Late in life he was given the task of compiling a register of shipping: perhaps an indication of the kind of task he did best.

Euclid Speidell had quite a wide circle of mathematical acquaintances, including such famous names as Edmond Halley. By the 1660s, in early middle age, he was in a position to vouch for the competence of other aspiring practitioners. He was, in other words, a respected and active member of the network of mathematical practitioners which spanned the British Isles – and far beyond – during this period. And he was able to draw directly on his family's tradition of mathematical teaching during his own mathematical career: he used his father's textbook, *An Arithmeticall Extraction*, to teach Nathaniel Denew in the 1680s. There were no copies of it available to purchase by that time, and Denew had to copy the book out by hand: it contained nearly 800 problems and answers. The experience prompted Euclid to bring the book back into print, taking the opportunity to add some notes about his father's method of teaching, perhaps in the hope of keeping alive his family's characteristic way of doing things.

His name, too, he kept alive. Euclid Speidell married twice, and was the father of at least ten children. Two were named Euclid, the first having died in childhood. And in the following generation, one of Euclid's grandchildren was christened Euclid in November 1695. But, despite all these efforts, the 'Euclid Speidell' brand did not really survive the death of its first holder, and the mathematical works of both John Speidell and his son fell into obscurity.

Nevertheless, the performance of the Speidell family illustrates that, thanks to the efforts of several generations of editors and practitioners, 'Euclid' was becoming a costume that almost anyone might wear. Far from being the exclusive property of learned scholars, editors or philosophers, the Euclidean persona could now be taken on even by a middle-ranking mathematical practitioner in Restoration London, writing in English and working as teacher, quantity surveyor and minor author.

Isaac Newton

Mathematical principles

The library of Trinity College, Cambridge: Christopher Wren's beautiful construction overlooking the River Cam. It stands on pillars against flooding, and the light catches it morning and afternoon. A bay at the south end of the reading room holds about half of the books that once belonged to Isaac Newton, most famous of the college's famous progeny.

One of them is a copy of Euclid's *Elements*, printed in Cambridge in 1655. It is small and fairly slight, about four inches by six. Bookplates pasted into the first few pages record the book's owners after Newton. Smudges and minor corrections indicate that the book was used, and not always carefully. In many of the margins Newton's neat, small handwriting is to be found: dozens of pages bear the notes he made while reading Euclid.

A number of Newton's books are marked with his manuscript marginalia, though few as much as this one. He most likely wrote in it during the 1660s, when he was first becoming seriously interested in mathematics as well as in natural philosophy.

It started in 1663 when, aged twenty, he went to Stourbridge Fair and bought himself a book of astrology, curious to see how the subject worked. After a certain amount of nonsense and a certain amount of sense (perhaps), he came to the construction of what was called a 'figure' of the heavens, and found he couldn't understand the explanation. He bought a book of trigonometry, hoping it would help. It did not, because it assumed a certain knowledge of geometry. Newton returned to the bookshop, and bought some geometry books.

If his later reminiscences are to be believed, a first glance through Euclid's *Elements* persuaded Newton that the book contained only trivialities, propositions too simple to need proofs: he was not the first to pass that judgement on the *Elements*' early propositions. He laid it aside and turned to the works of René Descartes, whose geometry was by reputation modern, exciting and difficult. Descartes had applied algebraic methods to geometry, describing curved lines using equations and solving geometrical problems by converting them first into their algebraic equivalents.

Newton absorbed this kind of thing rapidly, both from the writings of Descartes himself and from those of his followers. He picked up some peculiarly English algebraic notation from other books along the way. In several of his algebra books he left marginal annotations and notes, recording the progress of his thoughts. This was taking place mainly in his home town of Woolsthorpe during 1665–7, while the plague had effectively closed the University of Cambridge.

At last he turned back to Euclid's *Elements*, and he obtained a copy of the version edited by Isaac Barrow, a fellow of Newton's own college and the Lucasian Professor of Mathematics. Barrow was a conservative mathematician, who wished 'not to compose Elements of Geometry any-wise at my discretion, but to demonstrate Euclide himself, and all of him'. He was suspicious of Descartes' revolution and the practice of translating geometry wholesale into algebra. But in the interest of conciseness he was willing to use some algebraic symbols. Instead of writing out 'is equal to' he used an equals sign; similarly with addition, subtraction and other mathematical operations he used symbols rather than words. He boasted in a preface that no one could have rendered the proofs more concise than he had. The result was an edition of the *Elements* that looked rather different from any earlier one; it was also much shorter, fitting the whole fifteen books into well under 350 small pages.

Newton took Euclid's *Elements* much more seriously than he had on his first glance, and this time he annotated the book in some detail. His preferred way of doing so was, perhaps predictably, algebraically: he translated proposition after proposition from Barrow's hybrid language into fully algebraic terms. Where Barrow (and Euclid) wrote about points and lines, Newton wrote about numbers and variables: *x*s and *y*s.

Newton worked through four books of the *Elements* in particular, covering both geometry and the theory of ratios. His annotations in each book provided terse algebraic equivalents for Barrow's still quite wordy formulations. For the second proposition in Book 2, Barrow had a verbal explanation about dividing a given line into two parts and making rectangles and squares out of the parts. Newton wrote 'If $a = b + c$, then $aa = ab + ac$.'

Newton annotates Euclid.

There were dozens of similar cases, as Newton consolidated both his understanding of Euclid and his dexterity with algebraic language. By the later parts of the *Elements* the algebra itself attained a sometimes fantastic complexity, with multiple lines of symbols, square roots and brackets as well as notations Newton seems to have made up on the spot for certain types of relationship between ratios. But for Newton, this was evidently still a clearer and a simpler mathematical language than anything Barrow (or Euclid) had to offer.

It was quite normal for readers of mathematical textbooks to scribble in them, and indeed quite normal for them to consolidate their understanding of the contents by making their own version of it, whether a literal copy of a set of worked examples, or a translation into different terms or a different means of expression. Libraries hold thousands of copies of the *Elements* with readers' marks of just those kinds: margins filled with more or less fluent copies of the material printed on the pages, or translations of it from words into diagrams, from diagrams into numbers, or – as in Newton's case – from words into algebra.

But Newton worked with an extraordinary proficiency. There is something almost unnerving about the accuracy with which his annotations take convoluted verbal statements and render them into succinct and algebraic equivalents. Those equivalents are often much easier to understand for a modern reader, but Newton was working at speed in a novel mathematical language which he had taught himself as an undergraduate. His annotations are the record of a most remarkable mind at work.

Newton recalled his mathematical development, and his reading of Euclid, in autobiographical writings and reminiscences thirty years and more after they took place. The exact sequence of events is a little uncertain, and some of the information seems to have been filtered through the preconceptions of friends and admirers who were ready to be wildly impressed by any crumb from the great man's lips. Newton's work during these years has become the stuff of legend, and his intellectual heroism in working through Descartes, Euclid and other mathematical books unaided and alone has become part of his myth. In fact, even the year in which he annotated Isaac Barrow's edition of the *Elements* is not quite clear. But annotate it he certainly did; the book is still to be seen in his old college in Cambridge.

Later, of course, he would voyage still further on the seas of thought. It was in the same few years in the mid-1660s that he famously invented the calculus; over the next few decades he would work on problems in optics and the motion of the planets, bringing his matchless mathematical intelligence to bear and devising solutions that in some cases would endure for centuries. His 1687 *Principia mathematica*, indeed, was one of the great moments for the theorem-and-proof style of writing: the book adopted a superficially 'Euclidean' look throughout, with every idea relentlessly presented through rigorous deduction involving points, lines and ratios. On the surface, then, the book looked much as geometrical writing consistently had for 2,000 years, with lettered diagrams and wordy proofs that referred to the diagrams in detail. Each theorem had its proof; many had 'corollaries' or 'scholia' to add and explain details, just as in the writings of the Greek geometers.

Like others, Newton saw that by making a geometrical construction analogous – precisely, quantitatively analogous – to something in the real world, one could study the real world by studying the geometrical construction. Here is a geometrical diagram; here is how it matches what is going on in the solar system; here, then, is how the solar system works. This was a vital step in making geometry – or algebra – speak to the real world, making them tools for reading the book of nature. Its depth of application by him and the breadth of the fields to which his followers applied it made this *the* technique of the eighteenth-century Newtonian natural philosopher. Everything from the movement of the tides to the tuning of organ pipes, from determining the mass of the earth to the assessment of testimony in support of Christian theology, would sooner or later be subjected to such a Newtonian approach. Perhaps this was the last and most important child of the marriage of geometry and the practical disciplines – 'practical mathematics' – that had gathered pace since the fifteenth century.

On the other hand, the actual mathematics of Newton's book was also revolutionary in other ways, and went far beyond anything he could have found in the *Elements*. It dealt with quantities that started out finite but became vanishingly small and then became zero; it dealt with ratios of two such quantities as they *both* became equal to zero. Newton's mathematics was a descendant of Euclid's, but not one that Euclid himself would have recognised.

Interlude

By the eighteenth century, people had been reading Euclid's *Elements* for over 2,000 years. They read it, studied it, copied it, paraphrased or rewrote it. They translated it into different languages or different notations. They admired it, venerated it, attempted to cleanse and purify the text. They extended it, used it as a spur to their own mathematics, attempted to improve it, fill in its gaps. They used its structure as a model of knowledge, its arguments and its objects as an inspiration for philosophies. They used it to improve and train their minds, to change their own social status or identity. They used its methods and its conclusions as practical tools; they conflated its text and its methods with those of surveying or those of practical mathematics. They put it on stage; they made it into art and design.

The *Elements*' longevity was a result of the range of ways it could be read. Like almost no other book, it was devoid of instructions about what to do with it. There was no 'right' way to read the *Elements*, no orthodox interpretation. Generation after generation of readers had come to the text and done something with it that was meaningful to them. To them Euclid was author, sage and hero: ancient but always new; wise and powerful.

But then things began to go wrong.

IV

Shadow and Mask

Mary Fairfax

Euclid and the straitjacket

Burntisland, Scotland; around 1795.

I had to take part in the household affairs, and to make and mend my own clothes. I rose early, played on the piano, and painted during the time I could spare in the daylight hours, but I sat up very late reading Euclid. The servants, however, told my mother 'It was no wonder the stock of candles was soon exhausted, for Miss Mary sat up reading till a very late hour'; whereupon an order was given to take away my candle as soon as I was in bed. I had, however, already gone through the first six books of Euclid, and now I was thrown on my memory, which I exercised by beginning at the first book, and demonstrating in my mind a certain number of problems every night, till I could nearly go through the whole. My father came home for a short time, and, somehow or other, finding out what I was about, said to my mother, 'Peg, we must put a stop to this, or we shall have Mary in a straitjacket one of these days. There was X., who went raving mad about the longitude!'

The cultural cachet of Euclid rode high in Britain by the end of the eighteenth century, when Mary Fairfax – the teenage daughter of an unlearned naval family – valiantly attempted to educate herself in the subject. But in a place and time when education for women was controversial and, for many, frankly absurd or abhorrent, it is no great surprise that her attempts provoked opposition and ended with parental sanctions. Rather more unexpected, though, is the particular form their objections took: that the study of Euclid would end with her with a damaged mind and 'in a straitjacket'.

The contrary view still had many supporters; indeed, it remained most likely the dominant one: that studying geometry improved, disciplined and beautified the mind, fitting it for other studies and steadying the tendency of youth to inattention and flightiness. The English Neoplatonist Thomas Taylor (1758–1835) was, by his programme of translation and original writing, bringing into greater visibility in English the view that geometry was a privileged means of making contact with the eternal and the divine; in 1788–9 he brought out his translation of the commentary of Proclus on Euclid's *Elements*. The Romantic poets Blake, Shelley and Wordsworth – among many others – were admirers; Blake was said to have studied Euclid with Taylor himself. One of the outcomes was an acclaimed passage in Wordsworth's *Prelude* celebrating the capacity of geometrical study to overcome sorrowful or distracting feelings; first the poet told a tale of a castaway who had managed to save a geometry book from his shipwreck and would

> part from company and take this book . . .
> To spots remote, and draw his diagrams
> With a long staff upon the sand, and thus
> Did oft beguile his sorrow.

Wordsworth next stated that he himself had found similar comfort in geometrical study:

> mighty is the charm
> Of those abstractions to a mind beset
> With images and haunted by herself,
> And specially delightful unto me
> Was that clear synthesis built up aloft
> So gracefully . . .
> an independent world,
> Created out of pure intelligence.

Yet for many, the actual experience of studying Euclid was more like torture than anything else: students found it difficult, repetitive, unattractive, and of scant evident relevance to other school subjects or to real life. An epigram from 1780 quipped about the Euclidean proposition traditionally called the 'bridge of asses': 'If this be rightly called the bridge of asses, / He's not the fool that sticks but he that passes.' Across the North Sea, Swedish poet and composer Carl Mikael Bellman noted in his memoirs that 'My brain still winces, when I think of Euclid.' However much nature had been conquered by Euclidean science, human nature showed an enduring tendency to resist.

Of course, it was not novel in the eighteenth century – or any century – for schoolchildren to complain that their lessons were boring and useless. Nor was there any real novelty to educators' warnings about the dangers of an unbalanced curriculum. What came to prominence in the decades around 1800 was the specific claim that too much mathematics could make a person mad. A number of commentators argued that the study of mathematics was a powerful drug: correctly administered, to be sure, it could indeed improve the mind, even curing certain types of mental disturbance, but carelessly or excessively taken it would make one antisocial or ultimately insane. The trope stretched back at least to the Renaissance, when Albrecht Dürer's famous engraving of 'Melancholy' took the traditional iconography of geometry – the dividers, the polyhedron, the book – and made it a symbol of personal and social failure and its attendant mental malaise. Seventeenth-century gossip attributed the early death of Blaise Pascal to the excessive study of geometry.

The complaints were threefold. First, that the study of mathematics would make one antisocial. 'If at a ball, a supper, or a party of pleasure, a man were to be solving, in his own mind, a problem in Euclid, he would be a very bad companion, and make a very poor figure in that company': thus Lord Chesterfield to his son, writing in the 1740s.

Second, that mathematics by its abstraction and its impracticality would cut one off from human feeling, because 'primitive truths, those which are seized by feeling and genius, are not susceptible of demonstration' (thus Madame de Staël). In the world of Romantic literature, too much thought and too little feeling amounted to an unforgivable offence, and William Blake himself had announced that stamping mathematical laws onto the world was the work not of an inspired philosopher but of an evil demiurge. Precisely the dulling of emotion that made mathematics an escape from the overwrought imaginations of a Wordsworth made it harmful for many in the general population.

Third, that too much mathematics could, as a consequence, actually unbalance a mind. William Paley, placed top in the mathematical examinations at Cambridge in 1763 and subsequently an influential philosopher, was reported as saying that 'two or three' were 'cracked' by the Cambridge system every year: 'some of them go mad; others are reduced to such a state of debility, both of mind and body, that they are unfit for any thing during the rest of their lives'. Well-documented mathematics-induced breakdowns occurred from time to time throughout the nineteenth century, including those of Francis Galton in the 1840s and of James Maurice Wilson in 1859, each of whom recovered and went on to a distinguished career. An 1825 article on the 'Regrets of a Cantab' provided personal testimony from an individual who had been reduced to 'social incapacitation, low self-esteem, disillusionment and emotional numbness' as a result of his Cambridge mathematical training. This author, too, recovered, but retained his belief that it was a mistake to see mathematics as the only or the best form of reasoning.

Mary Fairfax's parents were thus not eccentric – were not necessarily even incorrect – to believe that the excessive study of mathematics could do serious damage to her mental health, with effects that might not necessarily be reversible. Mary's story was typical. She was the child of a naval officer, and her childhood was spent largely in the quiet seaport town of Burntisland, across the Forth from Edinburgh. Her mother 'seldom read anything but the Bible, sermons, and the newspaper', and Mary was mostly discouraged from reading or learning anything beyond the catechism. Her father's ill-advised attempts to direct her reading were no more successful than was a brief year at a local boarding school, and from her early teens her education was self-directed.

She indulged tastes for natural history, collecting shells and flowers and watching the stars, and put herself as far as possible through a course of reading she found described in Hester Chapone's *Letters on the Improvement of the Mind*. 'I was not a favourite with my family at that period of my life', she wrote, 'because I was reserved and unexpansive, in consequence of the silence I was obliged to observe on the subjects which interested me.' The village schoolmaster taught her some astronomy; an uncle taught her Latin; and there were lessons at the piano, at sewing, at dancing and at arithmetic.

Mary chafed bitterly at the limited intellectual activities she was being offered, but one day on a visit a friend showed her

a monthly magazine with coloured plates of ladies' dresses, charades, and puzzles. At the end of a page I read what appeared to me to be simply an arithmetical question; but on turning the page I was surprised to see strange-looking lines mixed with letters, chiefly X'es and Y's, and asked; 'What is that?' 'Oh,' said Miss Ogilvie, 'it is a kind of arithmetic: they call it Algebra; but I can tell you nothing about it.'

Another chance encounter – a conversation overheard at a painting class – told her that Euclid's *Elements* was recommended as 'the foundation not only of perspective, but of astronomy and all mechanical science'. For Mary to visit a bookseller and ask for Euclid was, she judged, impossible, but her brother's tutor was willing to purchase her elementary books on geometry and algebra from Edinburgh: he returned with an *Elements* and a copy of the *Introduction to Algebra* (1782) by the English teacher John Bonnycastle: by this tortuous route Mary Fairfax achieved the contact with mathematics she craved.

Mary's life, and her struggles to access education, would for some years continue to be those of a woman of her class and time: she spent winters in Edinburgh, but the rest of the year in Burntisland or visiting friends and relatives elsewhere in Scotland. In 1804, aged twenty-four, she married her cousin Samuel Grieg, a naval officer and diplomat based in London. In the capital Mary was able to pursue her studies in mathematics and in French, but she found no encouragement from her husband.

Grieg's early death, three years after their marriage, was from the point of view of Mary's intellect something of a release, and it left her with a small inheritance and sufficient family connections to enter the world of mathematical correspondence and education with more success. Academics at Edinburgh University were helpful, as was Mary's second husband William Somerville, an army doctor who positively encouraged her studies. She studied all the sciences, but her favourite remained mathematics, and she had soon left Euclid far behind in her pursuit of modern algebra and calculus.

Mary Fairfax.

There followed for Mary Somerville a most remarkable career of almost fifty years – based first in London and later in Italy – of scientific publication and popularisation. She ranged across astronomy, physics, meteorology and geography; her textbooks were widely successful and were translated into French, German and Italian, as well as being adopted for use at various universities. Honours and fellowships followed from more than a dozen scientific societies. On her death in 1872 the *Morning Post* of London recorded that 'whatever difficulty we might experience in the middle of the nineteenth century in choosing a king of science, there could be no question whatever as to the queen of science'. Evidently her study of Euclid had done her no lasting harm.

François Peyrard

Manuscript 190

It is around 1808, and François Peyrard has recently been sacked from his job at the École Polytechnique in Paris (which he helped to found); he is soon to be sacked from his job at the Lycée Bonaparte. He combines the roles of firebrand, revolutionary, scholar and teacher; he was once arrested on suspicion of terrorism. But not all of his activities are so dramatic. He is examining manuscripts brought to Paris from Rome in the wake of Napoleon's campaigns: manuscripts of Euclid's *Elements*, in fact, of which he is the author of a well-received translation.

Most of the manuscripts say just what he expects them to say: there are twenty-two of them, close to half the Euclidean evidence known in the world. But there is also one outlier. Compared with the normal text it lacks words, clauses and whole sentences. It lacks the addendum that Theon of Alexandria famously put into the book. Its scribe has noted in one place that the 'old version' of the book says such-and-such, but the new version something else. Peyrard is amazed and astonished at this unsuspected diversity in the Euclidean text.

François Peyrard lived the kind of life one could hardly make up. He was born in 1759 (or 1760) in what is now the Haute-Loire: a territory, it has sometimes been pointed out, as volcanic as the man. Orphaned during his school years, he enrolled in the infantry regiment the Gardes-Françaises, thus both escaping his family's threat of seminary and gaining a foothold in Paris. Not by accident, he had picked a corps famous at the time for its liberality, and he later boasted of spending much of his time in study, dispensed from his duties and not even required to wear his uniform. He attended lectures, haunted libraries and cultivated the savants of the city. He studied the classical languages, and took a particular interest in geometry and mathematics. By the late 1780s he was teaching mathematics privately, and he married in 1787.

It seems he throve during the disturbances of the following years. He early discovered an enthusiasm for republicanism and for the Revolution, and he became a member of the Society of the Friends of the Constitution (later the Club des Jacobins, whose motto was famously 'Live free or die'). He may have continued to teach; certainly he published short works on philosophy and history and by 1792 held administrative positions in Paris. He played a role in experimenting on the best shape for projectiles and surveying mining and weapons manufacture on behalf of the Committee for Public Safety.

In 1793 he was one of the authors of a project to reform public education. His learning and his conversation appear to have impressed the right people, and he became one of the group that founded the so-called Grandes Écoles, including the celebrated École Polytechnique, and recruited their first teachers. A brief reversal during 1795 was not enough to stop his rise: following

Jacobin rioting he and others were arrested as 'terrorists', but by his own account he shed 'tears of blood over the crimes committed in the name of liberty' and secured his release.

He was soon teaching at the École, as well as publishing its journal and overseeing its library. He published translations from the Renaissance polymath Heinrich Cornelius Agrippa and from Roman poet Horace (including the obscene passages suppressed in earlier versions) and built up the École's library from its few hundred volumes to over 10,000, establishing it as an important research facility. It was during this period that Peyrard made his first version of the *Elements* in French. Although the French education system – and the École Polytechnique in particular – was a beacon for modern mathematics and modern methods, Euclid still had his defenders as a necessary foundation. In a preface, Peyrard quoted the French mathematical historian Jean-Étienne Montucla, whose view was that though many had tried to improve on Euclid, many had failed.

Peyrard had hard words for other editors and translators, reckoning that the recent English versions of the *Elements* contained grave errors and the French versions in common circulation deviated from Euclid's text 'at every moment'. His own translation contained seven books, comprising the traditional school/university selection, minus some of the theory of ratios. It was faithful, literal and even word for word as far as possible. Like many other editors, he supplemented the *Elements* with an appendix containing material about the circle, cylinder, cone and sphere and about the area of surfaces and the volume of solids, derived ultimately from Archimedes. Citizens Lagrange and Delambre approved the book in Year 12 of the Revolution (1803/4), on behalf of the physical science and mathematics class of the Institut National; it was printed, and was adopted for teaching at the École Polytechnique and elsewhere.

There Peyrard's work on Euclid might easily have ended. His

was not an easy personality, and by the time his book appeared he had in fact been driven out of the École Polytechnique. He had contrived to fight with his students, his assistants and certain of his colleagues, and furthermore the director of the École had felt obliged to reprimand him for his irregular domestic arrangements (the complaint seems to have centred on loud altercations with his wife, together with the unconcealed presence of another woman in Peyrard's quarters). He was required to leave his quarters during July 1803, and towards the end of the following year the school, now with military status under the Napoleonic regime, dismissed him altogether. Much of what is known about his life derives from a lengthy memoir Peyrard wrote at this time in an attempt (on the whole unsuccessful) to justify himself. He was appointed a professor at the Lycée Bonaparte, also in Paris, where he continued his career of difficult conduct but also his work on Greek geometry.

He translated works of Archimedes, for publication in 1807, and he planned to work on both Apollonius and Diophantus. But the text of Euclid continued to hold his attention, and he became convinced that the old Oxford edition of the Greek text, which he had used as the basis of his translation, abridged or altered certain passages compared with Euclid's (presumed) original.

The exact chronology is nowhere made clear in Peyrard's writings, but it is certain that the French geometer Gaspard Monge travelled to Italy in Napoleon's wake and was able to send manuscripts on 'loan' from the major libraries there to Paris. (Napoleon was president of (northern) Italy from 1802, its king from 1805; the Papal States including Rome were annexed in 1809.) Peyrard was ultimately able to consult no fewer than twenty-three manuscripts of the *Elements*, ranging in date from the ninth to the sixteenth century.

It became clear that none of these manuscripts had been the

exact basis for the two previous printed editions of the whole Greek text: the rather faulty 1533 printing at Basel and Gregory's 1703 edition at Oxford. For all that, most of the manuscripts conformed rather closely to one another – and to the printed versions – in their texts. But one was different: from the Vatican library, designated number 190 in that library's catalogue of Greek manuscripts. It was from the ninth century (it is now dated to 830–50, making it older than the copy Stephanos made for Arethas of Patras, though Peyrard placed it closer to the end of the century), and it had not been used for a printed edition before. It contained *Elements* books 1–13, Euclid's *Data*, and then *Elements* books 14 and 15.

The text was not vastly different from the other manuscripts, but it was noticeable that it lacked some of their helping phrases, clarifications and additions. And it was particularly striking that it lacked the additional ending of proposition 26 of book 6: Theon's addition. Moreover, in the margin of manuscript 190, a hand different from that of the original scribe had subsequently added in that extra bit of text. Other marginal annotations noted that one proposition in book 13 was absent from most copies of 'the new edition but is found in those of the old'. The opposite was the case with one proposition in book 11.

Peyrard's excitement was intense, and he believed he had before him 'the pure text of Euclid', to be contrasted with and preferred to the edition of Theon which he surmised was represented by all the other sources. He worked steadily through the manuscript, marking up a copy of Gregory's edition with its variations. He compared other manuscripts, and considered each of the variations on its merits, trying to prefer manuscript 190 wherever he could. Thus he began to purge the Euclidean text of what he had become convinced were unwarranted accretions: alternative proofs, corollaries, lemmas and scholia.

His work was not perfectly consistent, and his faith in the text of manuscript 190 – or indeed of any manuscript – seems to have wavered at times; there were propositions for which he followed the printed editions despite the lack of any manuscript evidence, such as one proposition in book 10, which was not in manuscript 190 but which Peyrard did not want to remove, since to do so would have disturbed the numbering of the rest of the book. Elsewhere, he allowed his judgements about mathematical correctness and completeness to override the manuscript evidence, and a good number of lemmas and alternative proofs absent from manuscript 190 remained in his edition as a result. Occasionally he emended the text outright, when he found points at which neither the manuscripts nor the editions contained a version he thought correct.

By autumn 1809, Peyrard was in a position to turn his work over to colleagues for approval. To a committee consisting of the noted mathematicians Delambre, Lagrange and Legendre he sent both manuscript 190 and his marked-up copy of the Oxford edition; they certified the accuracy of his readings and urged that the work be finished. Subsequently Delambre and Gaspard Prony – who had seen the same material – reported to the Académie des Sciences that the new text seemed to them purer, clearer and more intelligible than the old. Their report hinted, though, that Peyrard's use of the manuscripts was turning out to be somewhat eclectic, and asked that if it was not possible to have an edition purged of all the faults that the manuscripts could correct, and enriched with all the additions they could furnish, he should at least provide a list of all the variant readings as an appendix to his edition.

From the Greek text he had now prepared, Peyrard made a Latin translation (of remarkable word-for-word clumsiness) and also marked up his French version of the *Elements* with the changes that the new Greek text necessitated. All of this took time, and he received permission directly from the minister of the interior to retain manuscript 190 until the whole edition was published (manuscripts from Italy were beginning to be returned to their owners by this time). Publication began in 1814, after the Académie had commissioned reports on the book from both its science committee and its history and ancient literature committee. The former was broadly favourable; the latter seemed less convinced that the new edition was a real improvement over the older ones, but the Académie granted its approval and printing took place. The three volumes appeared in 1814, 1816 and 1818. The fall of Napoleon and restoration of the Bourbon monarchy led Peyrard – who had already lived through more than his share of regime changes – to dedicate volumes 2 and 3 to the new king.

It was an impressive performance by any standard, with over 1,600 pages of Greek and Latin in two parallel columns, and a French translation at the foot of the page, as well as a critical apparatus listing how the printed text differed from manuscript 190 and from the Oxford edition, and notes on the faults in the 1533 Basel edition of the *Elements* as well.

Meanwhile, Peyrard's personal life continued to be difficult: his memoir, and the prefaces to his books, reported in snatches a dismal tale of the deaths of children and of at least one grandchild, and the flight of one of his daughters. At some stage he left (or was dismissed from) the Lycée Bonaparte, quite probably before his Euclidean edition was complete. He turned up in 1810 as a member of the minister of war's staff dealing with artillery; in 1816 he referred (in the preface to the final volume of the *Elements*) to sixteen years of 'calumnies' and 'persecutions', which does not suggest he had become any easier to work with. It is astonishing that he was able to complete his detailed textual work at all under these conditions, and although he planned to follow his Euclid with three-language critical editions of Archimedes and of Apollonius (the latter seems to have got as far as approval by the Académie des Sciences), it is no great surprise that they never saw the light of day.

François Peyrard died in 1822, in his early sixties, at the Hôpital Saint-Louis in Paris. The location implies destitution; the hospital was reserved for the needy, and it seems the great survivor had not survived the Bourbon restoration as well as he had once hoped.

⟨△⟩

Peyrard's discovery had raised the horrible possibility that the Euclidean text as it was generally known was deceptive: which hardly strengthened the case for a reverent attention to the exact words of its proofs in schools, universities, or anywhere else. But it had also raised the possibility of a purer, better, even a pristine Euclid, witnessed in part by manuscript 190 and – who knew – hiding in some unexplored archive or recoverable by a sufficiently diligent study of the manuscripts already known.

This was a period when a similar degree of suspicion was being brought to other ancient texts. Homeric scholars – both before and after the publication of Friedrich August Wolf's notorious *Prolegomena to Homer* in 1795 – were busily dividing up the *Iliad* and the *Odyssey* into their supposed component parts and attempting to date and locate those parts; the very existence of Homer himself was coming into question, and even if he existed many scholars would allow him no more than the function of a compiler. The Old Testament was beginning to go the same way, with a scholarly consensus eventually emerging around four different authors or schools of authorship for different parts of the text, the voice changing from chapter to chapter and even from verse to verse. For the New Testament, a fresh attention to manuscript evidence was producing a gradual change from the text received in print since the sixteenth century (derived from some quite late Byzantine sources) to a new scholarly reconstruction based on earlier and less Byzantine-coloured evidence.

The history of the Euclidean text after Peyrard defies easy summary, and attests to the degree of anxiety and suspicion – even despair – he had created. The edition by Danish philologist Johan Ludvig Heiberg in the 1880s attempted a more consistent version of Peyrard's approach, identifying the difference between manuscript 190 and the rest of the Greek manuscripts of the *Elements* with the difference between the pre-Theon and post-Theon versions of the text. Thus, Heiberg believed he could identify in some detail the changes Theon had made, and pass judgement on them. He studied a large proportion of the extant Greek evidence (today about eighty manuscripts containing all or part of the *Elements* have been catalogued, roughly thirty of them from before the fifteenth century), but not all would find his conclusions convincing.

Other historians pointed out that there is also substantial variation between the Greek texts – taken as a whole – and the Arabic and Latin versions of the *Elements* made in the Middle Ages; certain scholars pressed hard for *that* difference to reflect ultimately the difference between pre- and post-Theon, on the supposition that the early Arabic translations rested (as indeed they must have, given their date) on Greek manuscripts now lost. In this view, it was the Arabic and Latin translations which preserved something closest to the actual words of Euclid, largely lost in Greek over the centuries. In fact, a manuscript in Bologna turned out to contain, for parts of books 11 and 12, a Greek version of the *Elements* agreeing closely with certain of the Arabic versions.

The basic problem was that 1,200 years had elapsed between the composition of the *Elements* and the first complete version of the text to have survived into the modern age. Peyrard's discovery showed that by the end of that period there were (at least) two versions of the text in circulation, but beyond that it was extremely hard to peer into the abyss and deduce what had happened to the text: the Greek evidence pertaining to the gap amounted to one fragmentary palimpsest from the seventh or eighth century, a few papyrus fragments and a set of ostraka with a text so wild it was not even clear they were relevant to the transmission of the text as such: plus the knowledge that Theon made an edition, and his statement that he added a particular fragment to the text. Plus quotations in the early commentaries on the *Elements*, some of them preserved in Arabic translations.

New attention to the Arabic and Latin versions in the second half of the twentieth century made it clear, though, that those versions themselves showed substantial variation in books 1–10 (books 11–13 show a strikingly constant text), larger than the variation within the extant Greek materials: this despite the fact that the Arabic translators evidently took pains to consult multiple Greek manuscripts and to get the details right. Evidently, deliberate improvement of the text went on in Greek, Arabic and Latin over a period of at least a millennium and a half, attesting to the book's success and resulting in a range of different versions in circulation at any one time, from lean epitomes to teaching texts amplified with alternative proofs, extra propositions, explanations and remarks of every kind. At least some of what exists in the Greek manuscripts is almost certainly inauthentic, and there are almost certainly places where the Arabic and Latin versions do have an older form of the text: but the different versions have contaminated each other again and again, making certainty about details frequently impossible: and the pattern changes from book to book within the *Elements*.

With so much variation in view, and with no Greek evidence known that would determine with certainty when each part of that variation first arose, it now seems unlikely that the detailed history of the text will ever be reconstructed in full. The most recent comprehensive study specifically states that there would be no point making a new critical edition at this time, and it has been well said by a scholar of the Arabic *Elements* that the time for synthesis and certainty has not yet arrived.

Nicolai Ivanovich Lobachevskii

Parallels

Russia, under Tsar Nicholas I; 23 February 1826 in Kazan, on the Volga.

One mathematics professor attempts a revolution, of a kind. What if some of Euclid's assumptions were wrong? What if a new geometry could be created? What if space was not truly what we have all been assuming?

His lecture is duly noted, but his colleagues show no sign of caring.

△

The saga of Euclid's 'parallel postulate' had run and run: the complicated, non-obvious assumption about parallel lines that Levi ben Gershom, among others, had tried to prove from simpler principles. Ancient, medieval and Renaissance mathematicians had tried to replace it or to prove it, and all had failed. In the sixteenth and seventeenth centuries, when much attention was given to tidying up the 'fundamentals' of the *Elements*, a number of alternatives to the parallel postulate appeared, suggested in different editions. All failed to take hold: either they proved to be weaker and less useful than the parallel postulate, or they were not intuitive enough to be satisfactory as basic assumptions in a book of geometry.

By the eighteenth century, then, the parallel postulate had become a well-established fault in the status of Euclid's *Elements* as an axiomatic, logical treatment of geometry and space. Newton's reliance on Euclid's geometry meant that now much of natural philosophy embodied whatever it was that Euclid's basic assumptions meant: which added to their plausibility as a description of space, but also added to the degree of concern they attracted from mathematicians. Hardly a year passed without a new attempted proof being published for the parallel postulate; catalogues of authors and works list hundreds of items, from flimsy pamphlets to the 300 pages of Giovanni Girolamo Saccheri's *Euclid Freed of Every Flaw*, a rigorous working-out of the consequences if the postulate was *not* taken to be true. Saccheri thought those consequences self-evidently absurd, but others were, as usual, not convinced. By the end of the eighteenth century it was beginning to be the accepted view that the parallel postulate just could not be proved: that it, or an equivalent, would simply have to be accepted without proof.

Nicolai Ivanovich Lobachevskii was the son of a government employee; his parents may have been of Polish origin. Born in Nizhni Novgorod in 1792, he was enrolled in the high school at Kazan in or around 1802 on a public scholarship. He would spend much of the rest of his life in Kazan, a town of perhaps 40,000 or so at the confluence of the Volga and the Kazanka. The town was strictly speaking in 'European' Russia, but it stood more than 400 miles east of Moscow, well outside the orbit or the knowledge of most European intellectuals.

Happily for Lobachevskii, a university was founded at Kazan just a few years after his arrival. The high school itself dated only from 1798; Kazan University, initially administered as an adjunct to the school, opened its doors in 1805, becoming Russia's third university after those of Petersburg and Moscow. It served the vast areas of eastern Russia: the Volga, the Urals, Siberia and the Caucasus. Some called it the *ultima Musarum Thule*; the furthest end of the Muses' world.

Lobachevskii entered the university aged fourteen, his scholarship requiring him to teach there for at least six years after graduation. He was lucky in his teachers, who included four distinguished professors imported from Germany. The teacher of pure mathematics, Johann Christian Martin Bartels, had once taught Carl Friedrich Gauss: probably the greatest mathematician then living.

Lobachevskii acquired a reputation for rebellion at Kazan University, and Bartels had to lobby the other professors to have his bad conduct overlooked so that he could be awarded his degree. By 1812 he was teaching mathematics, and in 1814 he gained the title of professor. He made astronomical observations, and studied further with Bartels.

His career continued through committee work, deanship of his department and, from 1827, rectorship of the university. He was someone who had not crumbled during the official anti-intellectualism of the early 1820s, and he was involved in building work, library reform, and the founding of the university's scientific journal. He married in 1832, but his domestic life is poorly documented: reports of the number of his children vary from seven up to fifteen or even eighteen. By 1840 he was a councillor of state.

Retirement in 1846 was followed by what seems to have been a period of poor health and meagre circumstances. Some reports say Lobachevskii mismanaged his wife's inheritance. He farmed a small estate up the Volga, kept sheep, built a watermill. Late in life he lost his sight. He planted a grove of nut trees, but did not live to gather their fruit, dying in February 1856.

△

Nicolai Ivanovich Lobachevskii is remembered not for his long service to the University of Kazan, nor for his efforts with Merino sheep and nut trees, nor yet for his few publications on calculus and trigonometry or his occasional participation in astronomical work. His name is associated instead with the parallel postulate, and his work on geometry from 1815 until his death.

Like many another, he attempted a proof of the parallel postulate; his lecture notes for 1815–17 contain, indeed, several attempted proofs. But like all others he failed, and he became convinced that the postulate was not in fact a consequence of the others but must be treated as a separate independent assumption. Thus, as he put it, the theory of parallel lines was subject to a 'momentous gap'; it was 'incomplete', 'imperfect'. He determined to construct a new geometry without the postulate: the first new geometry since antiquity, a new candidate for the true description of physical space.

Equivalent to Euclid's parallel postulate was the assumption that only one line can be drawn through a given point that will be parallel to a given straight line. Lobachevskii began his revolutionary geometry with the contrary assumption: that more than one such line could always be drawn. Thus arose his mind-bending definition of parallels. Take a point and a line. Then all the lines which go through the point can be placed in two classes: those which cut the given line and those which don't. The lines on the boundary of the two classes will be called 'parallel' to the given line.

Perhaps it doesn't sound like much, but it opened up a world of counter-intuitive novelty. The first twenty-eight propositions of the *Elements*, the ones that did not depend on the parallel postulate, could be proved just as before in this new geometry. But after that, everything was different. Lobachevskii found himself in a world in which the angles in a triangle added up to less than two right angles, and their sum depended on the size of the triangle: in fact, the difference varied with the area of the triangle. Parallel lines were not equidistant; they got closer together, but more and more slowly, so that they never actually met. Two triangles with the same set of angles were necessarily the same size. It was possible to choose three points without them all lying on either one straight line or one circle.

Lobachevskii did much of this work using not the methods of Euclidean geometry but those of trigonometry. For flat triangles in a normal Euclidean plane, there were formulas relating the lengths of the sides to the angles between them, via trigonometrical functions like tangent, sine and cosine. There were equivalent formulas to describe triangles on the surface of a sphere, and these – the formulas of 'spherical trigonometry' – were perfectly well known to navigators and cartographers whose work took place on the roughly spherical surface of the earth. Lobachevskii was able to find equivalent formulas again, for triangles in his new, 'imaginary' geometry; they looked rather like the spherical ones, in fact, but with systematic changes reflecting the fact that his space was curved in a different way. (Being in Lobachevskii's space would be somewhat more like being on the surface of a saddle than being on the surface of a sphere; but really there is no surface in Euclidean space that possesses the true features of Lobachevskii's geometry.)

Lobachevskii found no inherent contradictions within the rules of his strange new world, and the fact that he could find a coherent system of formulas analogous to those of trigonometry strongly suggested his geometry was in fact a coherent, consistent one. He became confident that it was just as valid as Euclid's: merely different.

That meant that his new geometry was indeed a candidate to

be the true description of space: as was Euclid's geometry. Which was right? That would have to be found out not by intuition or pure reason – as many believed – but by experiment. The influential philosophy of Immanuel Kant said that ideas about space were, effectively, innate: they were objectively correct intuitions, not derived from experience. Lobachevskii – who was very probably acquainted with Kant's work – effectively said the opposite. He wrote of 'obscurity' in fundamental ideas about geometry and insisted that essential ideas of mathematics were received from the external world by the senses. Thus the parallel postulate, or Lobachevskii's alternative to it, would be for all practical purposes a physical law: and he proposed to test it by experiment.

But experiment failed. On the small scale it was impossible to tell the difference between Euclid's geometry and Lobachevskii's, just as no one could tell whether the earth was curved or flat by measuring a small garden. Lobachevskii's idea was to measure the parallax of stars: the apparent change in their position as the earth moved from one side of its orbit to the other. If space was Lobachevskiian, the parallax could not fall below a certain minimum even for the most distant star. But unfortunately, instrument error made the results inconclusive. The best Lobachevskii could say was that 'in triangles whose sides are attainable for our measurement, the sum of the three angles is not indeed different from the two right angles by the hundredth part of a second' (a second being a 3,600th of a degree). On larger scales and with more accurate instruments, it remained perfectly possible, then, that Lobachevskii's was the true description of space.

Lobachevskii was one of mathematics' few genuine revolutionaries, a man for whom that great classic the *Elements* was no more than a hint to do something entirely different. His geometry was novel and his ideas truly extraordinary. He seems to have been a successful and convincing teacher. But unfortunately, when it came to writing about his geometrical ideas, he was less able to convince, and his life – and perhaps this was part of the reason for his demoralised state in retirement – became a doleful tale, marked by repeated failures to persuade his colleagues to take his geometrical work seriously.

He published at first in the *Kazan University Messenger*: not a publication which was widely read. After first announcing his new geometry in a lecture in 1826, for which neither the text nor any contemporary reactions survive, he published four papers on the subject over the following decade, without eliciting a reaction. Having published in Russian so far, in 1837 he turned to French, publishing in a well-known European mathematics journal, but his paper was loaded with formulas, and wrongly assumed readers already knew his Russian work. It too sank almost without trace. An 1840 book in German and, in 1855–6 his last publication, titled *Pangéométrie*, in French and Russian, did no better. A reviewer for the Petersburg Academy of Sciences ridiculed Lobachevskii's ideas. The high status of Kant and of Euclid, and the fact that parallel-postulate-proving was becoming, like circle-squaring and angle-trisecting, the domain of eccentrics, may all have played a part.

Lobachevskii, furthermore, never visited European institutions, nor corresponded with mathematicians outside Russia. No student took his ideas forward. Thus, the man who had created a new world received virtually no recognition during his lifetime. The one exception was a corresponding membership of the Göttingen Society of Sciences, arranged by Gauss in 1842. Gauss was one of the very few who had read his work with understanding (he learned Russian in order to do so), and he himself had speculated about new non-Euclidean geometries, but he neither developed nor published anything on the subject.

Indeed, a handful of other mathematicians in the early nineteenth century would turn out to be working along the same lines as Lobachevskii. An uncle-and-nephew partnership in Germany – F. K. Schweikart and Franz Taurinus – and a father-and-son pair in Hungary – the Bolyais – both constructed new geometries in the years around 1820, and both corresponded with Gauss about the subject. The Bolyais' work was the more extensive, but they did not go quite so far as Lobachevskii. Their work, too, was virtually ignored in their lifetimes.

It was not until after Lobachevskii was dead that non-Euclidean geometry was taken up by other mathematicians. In 1858 an Italian study reformulated Lobachevskii's ideas in still more algebraic language than his; by 1870 there was a seminar on Lobachevskii's geometry at the University of Berlin. Bernhard Riemann, in the 1860s and later, was able to study curved spaces in general, of which Lobachevskii's and Euclid's were both special cases; he inaugurated a period when the study of geometries as mathematical systems was distinguished from geometry as the study of actual physical space. The subject subsequently merged with the study of other classes of mathematical abstractions that were gaining attention around the same time, and made contact with the long-running claim that mathematics should rely on abstract definitions and proofs only, setting aside common-sense intuitions about number, space and line. Euclid's fifth postulate had turned out to be an unprovable assumption which defined the character of one particular type of space: perhaps that of real space, perhaps not.

And then came Albert Einstein. The new geometries studied in the nineteenth century had been curved to the same degree everywhere (like a sphere). In Einstein's general theory of relativity, space was curved by a different amount at different points, depending on the presence of mass. Space in this view was non-Euclidean: in fact, it abandoned not just Euclid's fifth postulate but his fourth as well, which said that right angles were the same everywhere. When general relativity passed crucial experimental tests in the early twentieth century, there was no longer any denying that the world's geometry was different from Euclid's. Euclid's *Elements*, long-lived as it was, had proved to be only a temporary solution to the problem of describing real space.

And Nicolai Ivanovich Lobachevskii came to be celebrated as a pioneer; one Russian historian, writing around his bicentenary, wrote that 'there is a clear thread of development between Lobachevskii's geometry (a geometry of constant negative curvature) and the geometry of the general theory of relativity (a non-Euclidean geometry of variable curvature)'. His works were translated, collected, republished; non-Euclidean ideas were taken up in art and literature. Jeremy Gray, historian of modern mathematics, remarks 'it is unlikely that mathematicians have yet exhausted the richnesses of the world that Lobachevskii discovered in the relative isolation of Kazan'.

Maggie and Tom
The torture of the mind

The schoolroom at King's Lorton, in about 1830:

'I don't think I *am* well, Father,' said Tom; 'I wish you'd ask Mr Stelling not to let me do Euclid; it brings on the toothache, I think.'

(The toothache was the only malady to which Tom had ever been subject.)

'Euclid, my lad – why, what's that?' said Mr Tulliver.

'Oh, I don't know; it's definitions, and axioms, and triangles, and things. It's a book I've got to learn in – there's no sense in it.'

'Go, go!' said Mr Tulliver, reprovingly; 'You mustn't say so. You must learn what your master tells you. He knows what it's right for you to learn.'

Tom and Maggie Tulliver grew up at Dorlcote Mill in Lincolnshire, the children of Jeremy Tulliver the miller. Both had some aptitude for mathematics, but as was the custom in the Britain of their day, their educations diverged sharply.

Tom and Maggie.

Tom was quick with practical things: worms, fish, birds; gates, handles and padlocks. He could draw almost a perfect square without measuring, and could confidently, accurately estimate numbers of horses and the length of a playground. He had a knack for grasping concrete things the right way first time.

His father saw education largely in terms of profit and practical advantage: 'I wouldn't make a downright lawyer of the lad — I should be sorry for him to be a raskill — but a sort o' engineer, or a surveyor, or an auctioneer and vallyer . . . or one o' them smartish businesses as are all profits and no outlay, only for a big watch-chain and a high stool.' He placed Tom as a boarder with a Mr Stelling, an Oxford graduate of no great imagination. His inflexible approach to teaching consisted largely of Latin grammar and Euclid, to be memorised without, perhaps, very much emphasis on understanding. There was a leavening of history, essay-writing and arithmetic. Stelling's favoured metaphor for the mind was an agricultural one, with classics and geometry the cultivation which made it ready to receive almost any subsequent crop.

This treatment did not, however, suit Tom's cast of mind. Presented with two equal triangles he could grasp in a flash that they were equal; but the Euclidean demonstration of the fact eluded him completely. Both his prayers and his conversations with his father began to include the request for 'no more Euclid', a subject for which Tom conceived a hearty dislike. An attempt to start on algebra also had to be abandoned. Encounters with other boys were demoralising; a schoolmate called Philip blandly accepted that 'All gentlemen learn the same things', and positively enjoyed knowing what everybody else knew: Latin, Euclid and 'all those things'.

A classical education also turned out to be a poor fit for Tom's practical needs. His father the miller became bankrupt; and Tom, brought home from Stelling's, found in effect that he now needed to unlearn what little he had acquired of Latin and Euclid in order to prepare for an occupation. 'I don't like Latin and those things,' he lamented. 'I don't know what I could do with them unless I went as usher in a school; and I don't know them well enough for that! besides, I would as soon carry a pair of panniers. I don't want to be that sort of person. I should like to enter into some business where I can get on.'

Little could be clearer than Tom's example of the failure of a Euclidean education in the nineteenth century: or at least of the failure of that education once it had become detached from humanity, and indeed from common sense about the needs of different students and their differences from one another.

Tom's sister Maggie, by contrast, was emphatically not presented with the mathematical or literary classics: despite, ironically, possessing more aptitude for them than he. Access to such things – at least in the teacher Stelling's view of the world – depended on gender, and when she asked if he could teach her Tom's lessons he put her down brutally. Girls, he insisted, have 'a great deal of superficial cleverness; but they couldn't go far into anything. They're quick and shallow.'

On her visit to Tom's school, Maggie in fact found his Latin text intriguing and attractive, while Euclid put her off badly. A look into the book persuaded her that it was 'nonsense . . . and very ugly stuff; nobody need want to make it out', though she remained convinced that she *could* make it out, if she only had the chance to learn it systematically.

Later, after the wreck of their father's finances, Maggie was able to test her self-belief. Tom's schoolbooks came home: the Latin dictionary, grammar and texts, a logic book and the exasperating Euclid. She judged that 'Latin, Euclid, and Logic would surely be a considerable step in masculine wisdom – in that knowledge which made men contented, and even glad to live', and she dreamed of seeing herself admired for her intellectual attainments. She filled her hours with study, and knew, for a while, the triumph of finding that she was indeed able to learn what she had been told were exclusively masculine studies.

Yet she came to find her studies dry; to find the fruit of knowledge not just tough-skinned but positively bitter. Maggie was easily distracted away from the abstractions of Latin and geometry to the world of the physical and real: like her brother. In time, she put the schoolbooks aside and turned to other pursuits.

Maggie's and Tom's encounters with Euclid revealed their differences as well as the ways in which the education they each received failed to connect with their aptitudes. The dominance of Euclid in British mathematical education was at its height in the Victorian period, from the universities and dissenting academies to schools large and small, and to tutors like Stelling. In early Victorian England, despite growing doubts about the results, the majority of educators practically lined up to repeat what were by now old tropes about the use of Euclid in training the mind and disciplining understanding. In a period when 'enthusiasm' was often looked on with suspicion, the frankly unexciting drill of his geometrical propositions – like that of Latin grammar – could have positive value in controlling the young.

This was by contrast with other national traditions: in France, Germany and Italy, for instance, the range of geometrical textbooks had remained much wider, and it does not seem that Euclid was much studied in schools or universities during the eighteenth or nineteenth centuries. But in Britain, Euclid was now something of an old friend, the names of certain propositions virtually household words (the 'asses' bridge', the 'windmill'). His name and his methods continued to be invoked by writers on other sciences who wished to borrow his prestige. Poet Samuel Taylor Coleridge, writing his *Aids to Reflection* in the 1820s, claimed that his moral arguments enjoyed the solidity of geometry: 'we have begun, as in geometry, with defining our Terms; and we proceed, like the Geometricians, with stating our POSTULATES; the difference being, that the Postulates of Geometry no man can deny, those of Moral Science are such as no good man will deny'. In 1838, scientist William Whewell produced a textbook on mechanics and hydrostatics that likewise adopted the scheme of definitions, axioms and postulates to create the appearance of certainty for what he called his *Mechanical Euclid*: 'I have aimed at making it such a coherent system of reasoning, as that of which Euclid's name is become a synonym.'

That there was famously 'no royal road' to geometry also made it a social leveller for some commentators. Yet the values of abstraction, purity and apolitical certainty were not of obvious relevance for those who would spend their lives working at a trade. Some teachers and textbooks, in the tradition of practical mathematics, aimed to demystify Euclid, to teach geometry with an emphasis on its practical applications and through everyday examples and explanations. This was emphatically Euclid the engineer rather than Euclid the philosopher, with the values of ease, rapidity and utility. Others, like Stelling, stayed with a more austere presentation, whether or not it was what their students wanted or needed.

Euclid was also becoming, slowly, somewhat more accessible to women, beyond the realm of determined autodidacts like Mary Fairfax. Although geometry textbooks were neither aimed at women nor – with exceedingly rare exceptions – did they acknowledge that women might be among their readers, lectures on mathematics for women were provided, by mid-century, in such institutions as the Ladies' College in London or the Queen's College in Harley Street. One attendee was Marian Evans, the creator of the fictional Tom and Maggie Tulliver.

Evans was a provincial middle-class child like many of the characters in her novels, born and brought up in Warwickshire. Intelligent, sharp-witted and imaginative, she left formal education at seventeen after her mother's death and kept house for her father. When he died in 1849, she found herself independent at the age of twenty-nine, and, after a period in Geneva, settled in London as a journalist.

She had shown an interest in mathematics at school, and in Geneva she had written to friends that it was her habit to 'take a dose of mathematics every day to prevent my brain from becoming quite soft': mathematics here became a sort of self-medication against low spirits. Other friends and correspondents would remark on her knowledge of calculus and of the geometry of conic sections; she became a friend of Sofia Kovalevskaya, professor of mathematics at the University of Stockholm. She studied algebra books and took private lessons; and in her first year in London – 1851 – she attended the course of lectures at the Ladies' College on Mondays and Thursdays, half-joking that she must stint herself in white gloves and clean collars to pay for the ticket. Like Maggie Tulliver, she was attracted to the hard-skinned fruit of mathematics; but unlike Maggie she did not give up.

The lecturer, Francis Newman – younger brother of the future cardinal John Henry – was of a progressive school of geometry teaching. Although he concurred with the view that geometry was a valuable exercise for the mind, teaching logical thought and imparting a taste and a faculty for lucid thinking, he also felt that the works of Greek geometers were unfit for modern students. He provided an updated, rearranged geometry that – he thought – was easier and more attractive to students and provided a better grounding for modern mathematics, while retaining the advantages of an axioms-and-deduction method.

A few years later in 1859, Marian Evans shot to fame with the publication of her first novel *Adam Bede* under the pseudonym George Eliot. *The Mill on the Floss* was her second. Queen Victoria herself was an admirer, and by the publication of *Middlemarch* in 1871–2 Evans was widely acknowledged to be her country's greatest living novelist. Several of her novels reflected her interest in geometry – and in education and its failures. Adam Bede, for instance, was a mathematically minded carpenter, apt to calculate quantities of timber and estimates of length (he owed something to Evans' own father).

The Mill on the Floss, in which Maggie Tulliver shared certain of Evans' own characteristics, used geometrical teaching and learning to explore the ways education could fail: could fail in its Victorian aim of 'improvement' (of people and society); could fail to overcome character and circumstance; could fail an individual and his or her needs. Scornful about the traditional ways of teaching Euclid and the classics, Evans showed how ignorance and bad luck shaped the educations received by Tom and Maggie: an ignorance from which, within the confines of the novel, they could not escape. Evans herself could, of course, see further. Increasingly sceptical about supposedly unquestionable truths of any kind, Evans found in mathematics a way not to close down difficulties but to open them up, and to articulate her searching questions about society, knowledge and ignorance.

During the final year of her life, Evans told a friend she was studying conic sections every morning 'because she didn't want to lose the power of learning'. In her final novel, *Daniel Deronda*, she depicted a mathematics student at Cambridge: one who found there were better things to do with mathematics than pursue an out-of-date curriculum focussed on competitive examinations. This was the same line of thought as in *The Mill on the Floss*, but in an enlarged and matured form. 'Men may dream', she wrote, 'in demonstrations and cut out an illusory world in the shape of axioms, definitions, and propositions, with a final exclusion of fact signed QED.' But 'no formulas for thinking will save us mortals from mistake in our imperfect apprehension of the matter to be thought about.'

Marian Evans died in 1880, of a combination of a throat infection and kidney disease. Tom and Maggie Tulliver met their end in the 1840s. Tom had found some success and restored the family from his father's bankruptcy; Maggie had moved through a shift of perspective via an intense spirituality, a forbidden love and an abandoned elopement. Resolution came unexpectedly, when the Floss flooded; the briefly reconciled siblings drowned together after their boat capsized. Maggie's thirst for knowledge, which had taken her to encounter Euclid both in her brother's schoolroom and in his books, remained unsatisfied to the end.

Simson in Urdu

The Euclidean empire

1884 saw the printing of a compendium of Euclid's *Elements*, containing 275 propositions. Outwardly it looks like many others; it fills 212 pages, each about nine inches high.

The book is almost entirely in Urdu; it was printed in Mathra in Uttar Pradesh, and according to its English preface it was prepared for those studying for the Normal School Certificate and the Middle Class Vernacular Examinations in the North-Western Provinces and Oudh.

The preface also explains the use of Euclid for this purpose:

> it can hardly be necessary to defend ourselves for having made Euclid's *Elements* the basis of the work . . . notwithstanding the numerous attempts which have been made by our best modern geometers to find an appropriate substitute, the *Elements* of Euclid has ever held the chief place in our Universities and Colleges, and is never likely to be superseded.

Euclid in Urdu.

In nineteenth-century Britain, 'Euclid' meant mainly, though not exclusively, the popular English edition of Robert Simson (1756), professor of mathematics at Glasgow, and its many imitations and reworkings. These were 'school' editions of the *Elements*, prizing pedagogy, as it was variously conceived, over textual fidelity. They omitted parts, they rearranged the text and on occasion they paraphrased. They glossed, added cross-references and explained, selecting explanatory matter from the long tradition of Euclidean commentary, some of which went right back to the earliest Greek centuries, some of which originated in the Arabic or Latin phases of the text's journey. Of the authentic thirteen books of the *Elements*, Simson included just eight: the ones discussing two- and three-dimensional geometry, and the ones about ratios. Euclid's treatment of numbers was omitted altogether, and so was his discussion of the regular solids.

The edition was successful, and it was pervasive; Simson's *Elements* went through thirty editions, varying in their contents, their physical size and their price. By 1845, one admirer could write that 'no rival has ever yet risen up to dispute with Simson's Euclid the possession of the schools'. Two later popular versions of the *Elements* – by John Playfair in 1795 and by Isaac Todhunter in 1862 – were frankly based on Simson's. And it was not just Britons like Anne Lister, Mary Fairfax or Tom and Maggie Tulliver who read these versions of the *Elements*.

When the British East India Company seized Indian land and revenues from the 1750s onwards, and set itself to rule a continent, it found many things. One of them was a well-established intellectual tradition, with thriving patronage of science and technology under the Mughal emperors. The British presence included a substantial number of mathematically trained military engineers and surveyors, some of whom took a real if frequently naïve interest in Indian science, mathematics and astronomy; there resulted an ongoing attempt to bring Indian arts and sciences into a relationship with their Western counterparts. Misunderstanding and condescension played their predictable part, and on the whole the British persuaded themselves that, lacking the obvious signs of Euclidean, deductive reasoning, Indian mathematics and astronomy were not real mathematics and astronomy at all.

There was irony at every level here; Britons frequently contrived to misconstrue both the age and the content of the documents, artefacts and traditions they encountered in Indian mathematics and astronomy, as well as to miss entirely the fact that an Islamic culture had been bringing Euclidean mathematics to the Indian continent since the Middle Ages. As recently as the 1720s, for instance, Jagannatha Samrat had produced a Sanskrit version of the *Elements*, based on the Arabic Euclidean compendium by Nasir al-Din al-Tusi. Arabic and Persian translations of the *Elements* too, as well as commentaries and summaries, were there to be noticed; but the British did not notice them.

Thus, British educators saw the planting of Euclidean mathematics in India as a part – albeit a small part – of their colonial mission, and when devising curricula for Indian schools they regularly included the *Elements*. In the 1820s alone there are reports of Euclid being studied at schools in Calcutta, Delhi, Agra and Banaras, attainments ranging from a bare knowledge of book 1 through to boys who had studied the whole of the first six books (and were therefore about as well grounded in Euclid as any schoolboys anywhere).

A long-running debate in British circles concerned the use of English versus local vernaculars for instruction; in practice both were taking place, with consequences for the provision of schoolbooks. Books in English could be imported from Britain; books in local languages were typically translated and printed in India. In 1824 the General Committee of Public Instruction in Calcutta created a committee to render English mathematics books into Arabic, Persian, Urdu and Hindi versions, for instance. By the end of the nineteenth century, many thousands of copies of such books had been produced. There are reports of the *Elements* translated into Marathi and Gujarati, into Sanskrit, into Urdu and Hindi and Arabic, into Ooriya. Their sources were the popular English versions of the *Elements*: sometimes complete, sometimes excerpted, selected or combined in new ways. The first three books of Playfair's Euclid were included complete in the 1846 *Encyclopaedia Bengalensis*, with a facing-page translation in Bengali and other additions. An Urdu version, in the 1880s, cited Euclidean textbooks by no fewer than six British authors as its sources.

Indian voices, and the experiences of individual Indian students with Euclid, are frustratingly hard to find, but school records, reports and syllabi occasionally provide glimpses. In 1844 at the English department of the Calcutta madrasa:

> In the junior scholarship examination Abdool Luteef, in particular, passed a most creditable examination. He demonstrated accurately two propositions of the third Book of Euclid and deduced another two of moderate difficulty, besides answering correctly several questions in simple equations of algebra, and in fractions and proportions of arithmetic.

The Indian universities also routinely taught Euclid, and required a knowledge of the earlier books of the *Elements* for their entrance exams. In an 1855 scholarship examination paper set by Bengal University, out of nine questions on pure mathematics, three were Euclidean (the others dealt with algebra and plane trigonometry). The Euclidean questions concerned the construction of figures similar to one another, the angles in a network of intersecting lines, and the angles made by the intersection of planes in space. In the entrance examination for Calcutta University, knowledge of the first three books of the *Elements* was required; 'the student was permitted to give his answers in any one living language in which he was examined'.

Back in England, mathematician J. J. Sylvester would say in a polemical context that 'there are some who rank Euclid as second in sacredness to the Bible alone, and as one of the advanced outposts of the British Constitution'; such was indeed the case. It was not just to India but around the globe that Britain had by mid-century exported Euclid, an expression of cultural power and self-confidence by a society that saw itself as the heir to all that was ancient and Greek. (The fact that the *Elements* was of North African origin seems to have universally escaped attention.) In Australia, Africa and Asia – from Cape Town to Cook's Creek to Calcutta – the logical rigour of the *Elements* was used to discipline minds, just as the army and censuses disciplined bodies and trigonometrical surveys disciplined landscape.

Empires come and go. Historian Rimi Chatterjee has written evocatively of the physical detritus left by British education in India. 'There is a street in Calcutta that links the oldest and most venerable of the city's educational institutions, running south–north next to the lake called Gol Dighi.' It contains colleges – the Scottish Church College, Bethune College for Girls, Sanskrit College, Presidency College, the David Hare School – and it's called College Street. On College Street, as she reports,

> you will not be able to go ten feet without being accosted by the vendors of books. There are very few bookshops on College Street; instead, there are the kind of stalls that gave the world the term 'stationer'.

At first the stationers' carts seem to be crammed full with test papers, notebooks, guidebooks 'and other instruments necessary for the delicate operation of passing an examination'. But in fact there is a good chance of finding an old textbook too: a grammar from the 1810s, say, or an edition of Shelley from the 1850s. Or maybe a battered old edition of Euclid? Chatterjee doesn't mention him, but where 'the detritus of the Raj is bought and sold along with other useful titbits of later times', surely he has a place.

The Euclidean books printed for British India are today rather mysterious. Few survive of the thousands that were printed. Those that do survive are often copies that were donated to libraries – the British Library perhaps, or the Bodleian – at an early stage and show no evidence of having been used. Many other copies, of course, must have been used, but by exactly whom and for exactly what, and in quite what circumstances, is most uncertain. Research on the subject is at an early stage, and it will be many years, perhaps, before this story can be told in full.

His modern rivals

An Oxford college study, 1879.

> MINOS *sleeping: to him enter the Phantasm of* EUCLID. MINOS *opens his eyes and regards him with a blank and stony gaze, without betraying the slightest surprise or even interest.*
>
> EUCLID: Now what is it you really require in a Manual of Geometry?
>
> MINOS: Excuse me, but – with all respect to a shade whose name I have reverenced from early boyhood – is not that *rather* an abrupt way of starting a conversation? Remember, we are twenty centuries apart in history, and consequently have never had a personal interview till now. Surely a few preliminary remarks—
>
> EUCLID: Centuries are long, my good sir, but *my* time to-night is short: and I never was a man of many words. So kindly waive all ceremony and answer my question.

No tradition can last forever. The *Elements* had been criticised for driving students mad, for torturing minds and, increasingly, for sheer irrelevance to modern mathematics. A culture that was increasingly apt to be suspicious of ancient texts, authors and authorities was becoming unmistakably suspicious of Euclid. The chorus of voices against him became overwhelming in the later decades of the nineteenth century.

The sheer popularity of Euclid in English had become a problem, as it made for a proliferation of different versions of the *Elements*, each with its own peculiarities of arrangement, presentation and notation; even the differences between different printings of the popular versions could prove confusing. No English edition in common circulation gave 'Euclid pure and simple'; few included the books on number theory or the regular solids; many modified the contents in detail by subtracting difficult parts or adding extras on subjects such as trigonometry. If the point was to ensure that everyone was learning the same thing in the same order, from a standard textbook, that point was being missed.

It did not help, either, that the nineteenth century was a period of anxiety for British mathematicians, who came to feel that they had fallen behind their continental counterparts by an over-reliance on Newton's methods and notation. A group calling itself the Analytical Society, as well as others, vigorously promoted the reform of university teaching and of the mathematical research agenda from the 1800s onwards. Omnipresent Euclid was, arguably, part of the problem, and duly came under attack.

Outspoken British opponents of Euclid included mathematician J. J. Sylvester, who announced himself vehemently as a 'hater of geometry' (meaning Euclid). He wished to see the *Elements* 'honourably shelved or buried "deeper than did ever plummet sound" out of the schoolboy's reach'. A widely quoted French report on the English education system condemned the study of Euclid as little better than rote learning, privileging memory over intelligence. By 1870 an 'anti-Euclid' organisation was suggested in a letter to *Nature*, and although what came of the proposal was called the Association for the Improvement of Geometrical Teaching – later the Mathematical Association – it was in its early years an anti-Euclid lobby in all but name. It issued annual reports and produced a geometry syllabus which it circulated to British universities and government departments.

It seemed that the traditional characteristics of the *Elements* – its beauty, wisdom and usefulness – had abruptly turned into their opposites. Suddenly it appeared a dead thing, textually corrupt, dry, rebarbative and maddening; blind to the realities of pedagogy, of modern mathematics, of mathematical applications in the industrial world.

But as well as critics, Euclid found some heavyweight defenders among British mathematicians. Augustus De Morgan and later Arthur Cayley, perhaps the first internationally important geometer England had produced since Newton, were both ardent enthusiasts. De Morgan wrote that 'There never has been, and till we see it we never shall believe that there can be, a system of geometry worthy of the name, which has any material departures (we do not speak of corrections or extensions or developments) from the plan laid down by Euclid.' Another supporter, Lewis Carroll, achieved in his *Euclid and His Modern Rivals* a book-length exposé of the shortcomings of the new textbooks that was in places actually funny, showing an Oxford tutor named Minos in dialogue with the ghost of departed Euclid about the confusions and absurdities they were easily able to identify in the latter's upstart rivals.

The arguments in Euclid's favour were pragmatic – it was better to have one textbook, one sequence of propositions, than many – and were added to the personal testimony of teachers who had used the *Elements* with success. The project to write something better appeared to be really failing, and this first round of the 'Euclid debate' was won by his defenders.

There was more to come. As well as the advantages of a single sequence of geometrical propositions, a point often argued in its favour was the status of the *Elements* as a unique, correct description of the real world: of space, lines, points and shapes. As Lobachevskii's non-Euclidean geometry became known in the English-speaking world during the 1870s, that ceased to be a tenable position. As William Kingdom Clifford put it in one of his essays on the subject, 'the geometer of to-day knows nothing about the nature of . . . space at an infinite distance. He knows nothing about the properties of this present space in a past or a future eternity.' Not just the usefulness but the truth of Euclid's system was now in question, and that novel fact did his cause no good at all.

In a further blow, the years around 1900 saw mathematicians, on the whole, become increasingly interested in the logical foundations of their subject; a move to more philosophically and logically grounded forms, with less reliance on intuition and several attempts to remake mathematics on completely logical, completely algebraic or symbolic foundations. An upshot was several attempts to measure Euclid's *Elements* against the standards of modern logic. It was inevitably found wanting. For millennia, indeed, commentators had been spotting odd gaps in the logical structure of the *Elements*: results that were never subsequently used, missing definitions, unstated assumptions. Suddenly they mattered. The great logician Bertrand Russell (1872–1970) wrote scathingly in 1902, in a paper specifically about the desirability of using Euclid in teaching, that 'his definitions do not always define, his axioms are not always indemonstrable, his demonstrations require many axioms of which he is quite unconscious', and thus that the value of the *Elements* as logical training had been 'grossly exaggerated'.

One proof early in the book Russell judged 'so bad that he would have done better to assume this proposition as an axiom'. (The complaints were not just of their time: recently Ian Mueller has written about the logical structure of the *Elements*, and has repeatedly drawn attention to similar matters: 'It is difficult to see that there is any proof here at all'; 'needlessly complex'; 'perplexing . . . from a logical point of view'; 'notoriously inadequate'.)

In a way this merely showed that once one had decided to treat a text with suspicion, it was easy to find things wrong with it. But the consequences for Euclid's status were real, and they did no good to those who still wished to see him as the natural foundation of a mathematical education and a complete and rigorous training in the ways of correct thought. The repeated statement that the *Elements* was not just a body of facts but a method of proof, a style of thought, the best possible training for the mind, was, like other optimistic traditional claims about Euclid, suddenly no longer viable.

The eventual outcome was still no certainty. No new textbook had succeeded in supplanting Euclid, and if a substantial amount of anxiety had been aired about the *Elements* and its usefulness as a textbook in the Victorian age, it would take more than that to remove him from schools and universities.

The final break came in the years from 1901, when education reformer John Perry brought his agenda to prominence. Perry opposed abstract reasoning, and particularly opposed its use in education. His view and his experience was that 'it is a very exceptional mind, and not, perhaps, a very healthy mind, which can learn things or train itself through abstract reasoning; nor, indeed, is much ever learnt in this way'. He argued forcefully for the use of experiment and intuition in the learning process, methods which he had used with success in his own practice as a teacher.

Perry could have been no more than another briefly prominent eccentric, but in 1901 he persuaded the influential British Association for the Advancement of Science to look into the reform of geometry teaching. Its report recommended both a flexibility in the choice of textbooks and the introduction of Perry's favoured methods of practical, experimental work, trial-and-error and learning-by-doing into geometry teaching.

Euclid's official demotion followed with surprising swiftness. The University of Cambridge commissioned a committee report on its geometry examinations. With Perry's ideas now in the mainstream, and an influential member of the Association on the committee, it was predictable that the committee would find against Euclid. It recommended that in university examinations any systematic proof, Euclidean or otherwise, be accepted for the propositions proposed. Adopted by the university's Senate, and coming into force during 1904, this effectively meant that learning Euclid was no longer compulsory at Cambridge.

Examining boards elsewhere in Britain were quick to follow, as were schools throughout the country and the British dominions overseas. From being geometry's king, Euclid had abruptly become one citizen among many. The *Elements* continued to be taught, and Euclidean methods to be learned and held up to admiration: but they would never again enjoy the exclusive sway that had once seemed their natural right.

Thomas Little Heath
The true con amore spirit

1908. A middle-aged civil servant is at home in London. In a large room, both music room and library, the man moves to and fro between desk and piano. He writes, fluently, in longhand, for hours at a time, but now and again he comes to a knot he cannot untie. Then he turns to the piano, and plays – Brahms, Schubert, Bach or Beethoven – until the knot is gone and he can see his way clear again. Then back to the desk, to write again.

Sir Thomas Little Heath (1861–1940) was a most unlikely hero for Euclid and his *Elements*. The son of a Lincolnshire farmer, his first teacher a self-educated tanner's son, Heath early committed himself to the values of hard work and self-discipline. By eighteen he was a student at Trinity College, Cambridge, and neither the sporting field nor the debating union lured him; his only reported activity besides work was music. As well as taking honours in both classics and mathematics (first class in the former; placed twelfth

Thomas Little Heath.

in the university in the latter) he learned modern languages: French, German, Italian and perhaps others.

In 1884, straight out of university, he passed first in the Civil Service examination and entered the Treasury as a clerk. The Treasury was a congenial, cultured environment, staffed in the main by Oxford and Cambridge graduates, and with none of the emphasis on technical economic competence that would later be expected. Most of its business involved handling applications from other Civil Service departments in connection with expenditure, and its main task amounted to parsimony. Prime Minister William Gladstone had memorably asserted that 'the saving of candle-ends' was 'very much the measure of a good Secretary of the Treasury'.

On-the-job training meant that junior officials moved from division to division every few years to give them a broader experience of the Treasury's work. Within a picture of broad stasis, it was growing; by 1914 it had more than thirty senior staff, and it was coming to be recognised as the de facto chief of the Civil Service departments, its head in practice the head of the Service.

Heath flourished there, moving up the ranks, and his loyal service was rewarded with honours. In 1903 he was made a Companion of the Bath and in 1909 a Knight Commander of the Bath. In 1913, aged fifty-two, he was joint permanent secretary, that is joint head of the Treasury, and thus at the effective pinnacle of the British Civil Service.

But everything would change over the next five years. Fiscal radicalism became the new norm under Lloyd George as prime minister, with war loans, land tax and super-tax. Public spending – and the national debt – ballooned, and Treasury control was greatly weakened: by 1916 the war departments were spending £2.5 million per day, and Britain was lending her allies nearly another £1 million daily as well. The Treasury was overworked, and the burdens on the permanent secretaries were huge.

The other permanent secretary, a man named Bradbury, handled finance (although he was notoriously error-prone in arithmetic), while Heath dealt with administration and the control of expenditure. In 1916 he was rewarded for his efforts with investiture as Knight Commander of the Royal Victorian Order. But in fact the strain was beginning to tell. He acquired a reputation for inflexibility: 'the special incarnation of all that is most angular and pedantic' according to opposition leader H. H. Asquith. In one incident he opposed introducing telephones into the Treasury, arguing that they would lead officials to neglect the art of writing concisely.

In the post-war reorganisation of the Treasury, Heath lost his position. His colleague Bradbury was moved to the Reparation Commission, and Heath to the National Debt Office, while a much younger man was put in at the head of the Treasury. Heath lingered for a time, as comptroller general and secretary to the Commissioners for the Reduction of the National Debt; he retired in 1926, aged sixty-five.

$$\triangle$$

It was not unknown for the cultured inhabitants of the Treasury to develop learned hobbies. A previous joint permanent secretary had pursued the study of Pali. Heath's hobby was ancient Greek mathematics.

He brought to that subject the same talent for meticulous, dogged work that had attracted remark in Cambridge and at the Treasury. In the years from 1885 to his death in 1940, Heath published around 5,000 printed pages, including English versions of the works of Diophantus, Apollonius, Archimedes and Euclid, a two-volume history of Greek mathematics, a one-volume account of Greek astronomy and a volume on mathematics in Aristotle, as well as about a dozen separate articles and two short popular books. Contemporaries reckoned it all 'sound and careful to the last degree', and ranked him as one of the half-dozen top historians of ancient mathematics in the world: 'one of the most learned and industrious scholars of our time'. He was awarded honorary degrees by Oxford and other universities, and became a fellow of both the Royal Society and the British Academy.

Heath's English version of the *Elements* appeared in 1908. His was a literal translation and a complete one: the first new version in English of certain parts of the text since the 1780s. His translation followed the edition of the text by Johan Heiberg, and therefore took up Heiberg's (and Peyrard's) belief that by following the now-famous manuscript 190 it was possible to obtain 'the genuine text of Euclid' as opposed to the edition of Theon. Heath was guided throughout his labours by Heiberg's assumptions about the ancient authors, and by the slant he had sometimes put on them in his decisions about what what was authentic and what was not, and how to present the texts and their accompanying diagrams. Archimedes, in this view, was precise but difficult; Apollonius was just plain difficult; but Euclid had to be lucid, accessible and even easy. Explanation was relegated to notes and footnotes, where Heath included wide-ranging information about Euclid's mathematics, its history and the previous editions of the text. His note on Pythagoras' theorem, for instance, filled eighteen pages of tiny print, and ranged across the Greek and Roman evidence for the antiquity and discovery of the theorem, with speculation as to the method of its first proof; discussed the concept of 'irrationality' and pairs of square numbers whose sum was another square number, including general rules for finding such triples and speculations as to how these were discovered; gave various strategies for the proof of the theorem with related examples from China and (at length) India, and finally set out certain ancient Greek extensions of the theorem. As well as sources in Greek and Latin, this particular note cited historical writing in both French and German, and there lurked in the background work that Heiberg had published in Danish and Latin. Further remarks in an appendix discussed the theorem's name, with citations from various languages plus a splendidly incongruous personal touch:

I venture to suggest that light may be thrown on the question by a very modern version of the 'Bride's Chair' which appeared during or since the War in *La Vie Parisienne*. The illustration represents Euclid's figure for I. 47 [Pythagoras' theorem] and, drawn over it, as on a frame, a *poilu* in full fighting kit carrying on his back his bride and his household belongings. Roughly speaking, the soldier is standing (or rather walking) in the middle of the large square, his head and shoulders are bending to the right within the contour of one of the small squares, while the lady, with mirror and powder-puff in action, is sitting with her back to him in the right angle between the two smaller squares.

(The copy of *La Vie Parisienne* in question has so far eluded rediscovery.)

In his notes, Heath adopted the view that Greek mathematics could in some instances be helpfully translated into modern algebra. Translation from one mathematical language to another to aid comprehension was not new: there were numerical glosses in the margins of the Byzantine manuscripts as well as a Byzantine translation of part of book 2 into arithmetical terms. Printed editions had rewritten Euclid as algebra since the seventeenth century, but there was a particular vogue from about the 1870s for arguing that the content of, say, Euclid book 2 *really was* algebra, effectively disguised beneath a geometrical language. Heath adopted this view whole-heartedly in his earlier books, but toned it down somewhat for his Euclid: 'The algebraical method has been preferred to Euclid's by some English editors; but it should not find favour with those who wish to preserve the essential features of Greek geometry as presented by its greatest exponents, or to appreciate their point of view.' The algebra was still there, but it was now placed in the notes, not the main text.

After more than a century, Heath's *Elements* has not faded from view; there was a second edition in 1926, and its main text can now be obtained in various printed versions including at least one Great Books series, as well as online. It supplanted its predecessors to become the standard English version of the book. One of Heath's motives for translating the *Elements* was to combat what he felt was the disastrous turn of events in Britain's abandonment of Euclid for teaching purposes. In 1908 he wrote that he hoped it would soon be possible to return to Euclid complete in teaching; he subsequently issued book 1 in Greek with that aim in mind. The second edition reprinted the same remarks, including the insistence that 'The body of doctrine contained in the recent textbooks of elementary geometry does not . . . show any substantial differences from that set forth in the Elements'; and that 'in the centuries which have since elapsed, there has been no need to reconstruct, still less to reject as unsound, any essential part of their doctrine'. But by 1926 the case was hopeless, and Heath's words had taken on the tone of a despairing appeal to the old image of an eternal, unchanging world of Euclidean certainty: 'Euclid can never at any time be more than apparently in abeyance; he is immortal.' A final blast from Heath's trumpet came in 1932, when he contributed a preface to a reprint of a Victorian edition: 'no alternative to the Elements has yet been produced', he wrote, 'which is open to fewer or less serious objections'; students should 'be introduced to Euclid in his original form as an antidote to the more or less feeble echoes of him that are to be found in the ordinary school textbooks of "geometry"'.

Heath believed not just that Euclid was fundamental to pedagogy but that the Greek achievement was fundamental to mathematics; that 'Mathematics in short is a Greek science', that 'Elementary geometry is Euclid'. He seems to have subscribed without much hesitation to the view that the ancient Greek cultural achievements were in various ways absolutely exceptional, and to the concomitant that all other ancient mathematical traditions were inferior if not insignificant. It is no surprise that he concurred with Heiberg's opinion that the Arabic evidence for the Euclidean text could safely be dismissed in most if not all cases.

He believed, further, that the study of the 'Greek genius' was an unmissable part of culture. 'Generation after generation of men and women will still have to go to school to the Greeks for the things in which they are our masters; and for this purpose they must continue to learn Greek.' And again, 'If one would understand the Greek genius fully, it would be a good plan to begin with their geometry.' The Elements, then, was 'one of the noblest monuments of antiquity': pure, noble and unchanging. Like those who presented the *Elements* in the British colonies, Heath evidently took Euclidean geometry as a sign of cultural superiority, and as part of his own cultural identity. The work of presenting it to a new generation, and of performing exegesis on what he persistently called its 'body of doctrine', seemed to him little less than a sacred duty.

Thomas Heath was an exceptionally able scholar, with a command of ancient Greek mathematics that would have been impressive in any period, including in ancient Greece itself. His love for Euclid's *Elements* – for a version of Euclid's *Elements* that was in part his own creation – was perfectly sincere, and it gave him a perspective on the text that was, perhaps, never to be repeated. He once suggested that 'any intelligent person with a fair recollection of schoolwork in elementary geometry would find it . . . easy reading, and should feel a real thrill in following its development'. 'I should be surprised', he wrote, if 'qualified readers, making the acquaintance of Euclid for the first time, did not find it fascinating, a book to be read in bed or on a holiday, a book as difficult as any detective story to lay down when once begun.' If this should seem improbable, he added that 'I know of one actual case, that of an undergraduate at Cambridge suddenly presented with a copy of Euclid, where this happened.' The specific example cannot now be identified; it is hard not to suspect that the only such case, ever, was that of Thomas Little Heath.

Heath had lived through a period when feeling against Euclid had been expressed with startling intensity, and when the *Elements* had lost, apparently forever, its place in British education. That he had experienced the debate about Euclid and its effects at first hand certainly coloured his willingness to express reverence, respect, and even love for the text. One reviewer commented specifically on this feature of his work, relating it to the traditional, vanishing, dependence of British education on Euclid: Heath's *Elements* was 'a work that no one but an Englishman could write in the true con amore spirit'.

There had long been, and for all his precipitous decline there would continue to be those who, like Heath, fell in love with Euclid. Thomas Hobbes in the seventeenth century was reported to be 'in love with geometry' following his chance encounter with the forty-seventh proposition. In the same century George Wharton, introducing an English edition, admired the book and divinised the man: 'Behold Great Euclid! But behold Him well! For 'tis in Him Divinity doth dwell!' In 1920s America, the author Edna St Vincent Millay produced perhaps the most memorable expression of that romantic love in her sonnet:

> Euclid alone has looked on Beauty bare.
> Let all who prate of Beauty hold their peace,
> And lay them prone upon the earth and cease
> To ponder on themselves, the while they stare
> At nothing, intricately drawn nowhere
> In shapes of shifting lineage; let geese
> Gabble and hiss, but heroes seek release
> From dusty bondage into luminous air.
> O blinding hour, O holy, terrible day,
> When first the shaft into his vision shone
> Of light anatomized! Euclid alone
> Has looked on Beauty bare. Fortunate they
> Who, though once only and then but far away,
> Have heard her massive sandal set on stone.

Max Ernst

Euclid's mask

May 1945; 42 East 57th Street, Manhattan: the Julien Levy Gallery.

The new exhibition is of paintings and sculptures by German artist-in-exile Max Ernst. One of the paintings bears the title *Euclid*.

It is an oil on canvas, nearly square: just over two feet high and a little less wide. The signature at the bottom right reads 'max ernst 45'. It is a portrait (is it a portrait?). It shows a broad-shouldered figure from the bust upwards (or does it?): torso towards the viewer, head turned slightly to its left.

Starting at the top, the eye meets a black feather (perhaps). Next, two white roses; but some viewers see an owl's face instead. These sit on a black hat with something of the turban to it. Below the hat, a pyramid with its point at the bottom: the pyramid's 'base' (at the top) is most likely a narrow rectangle, but it is covered by the hat. On the front face of the pyramid, a stylised leaf makes a sort of nose, suggesting a face, with two holes for eyes.

Below the face there is a mass of black drapery: perhaps it is fur, perhaps fabric. Upon it rests a white gardenia, that could instead be a shuttlecock. A grey jacket shows a single button and, on the right breast, a fish, perhaps in appliqué. There is the hint

Max Ernst's *Euclid*.

of a right arm. The left hand (is it a hand? There seem to be only three fingers, maybe four), gloved in white, holds another fish, this one looking less like part of the jacket. The fish curls around one of the fingers like an oversized ring.

The remainder of the canvas is filled with a chequered pattern made from diagonal lines: neither perfectly straight nor perfectly parallel and perpendicular to one another; towards the right they make rectangles and near-squares, but towards the left the pattern breaks down into wider curves. The shapes are coloured in cool blues, greens, greys and a cold yellow that echoes the figure's 'eyes'. There are a few white shapes, and two red. The chequers are not really a background, because there are three places at least where they cut in front of the hatted, jacketed figure. One of their bolder curves grows continuously out of the pattern on the jacket. Three of their straight lines continue onto the two sides of the pyramid/head that can be seen.

Max Ernst (1891–1976), born in Germany, was a naturalised citizen first of America, then of France. He assimilated the spirit and the visual language of the German Romantics, the Expressionists, the Primitivists and the Dadaists; for a decade and a half he was a declared Surrealist. An inexhaustible inventor of new techniques and styles, by his American days he was putting some distance between himself and doctrinaire Surrealism, with, on the one hand, a series of despairing works remembering a Europe dismembered by war, and on the other a series of apparently calmer, harmonious pieces of which *Euclid* was one. Like many of his works, though, *Euclid* defies easy interpretation. Although it has been called a high point of Surrealism, it has left its interpreters often helpless. An early viewer saw the chessboard of Lewis Carroll's *Through the Looking Glass* in the roughly squared background (Ernst did indeed love the Alice books, and would later illustrate both Carroll's *Hunting of the Snark* and his *Logic*). Another commentator wondered whether the rose symbolised alchemy or mysticism; and it has been suggested that the fish creates a pun, being 'out of its element'.

Certainly the traditional attributes of Euclidean portraiture are all absent. There is no compass, no rule, no book, no diagram and no students. Nor is there a trace of the classic geometer's pose, leaning over a diagram. Almost the only feature in common with the familiar depictions of Euclid from the Middle Ages and Renaissance onwards is the turban-like hat. The background's collapse of squares and rectangles into ambiguous intersecting curves has a whiff of non-Euclidean geometry to it. The clotted folded fabric of the wrap is the opposite of a Euclidean plane. And the ambiguous, not to say fragmented, perspective, in which foreground and background seem to link and overlap, seems, also, positively anti-Euclidean.

At the same time, recognisable geometrical forms are conspicuous by their absence: in contrast to much of twentieth-century art. One exception may be the fish, which possibly recalls the shape of a distorted torus sometimes featured in sets of geometrical teaching models (Man Ray had published photographs of such things from the Institut Henri Poincaré in Paris, which Ernst quite likely saw). Another is the pyramid that replaces the figure's head. Like the sixteenth-century artist Giuseppe Arcimboldo, whom the Surrealists loved, and who painted heads and figures built out of flowers or fruits, or a librarian made out of books, Ernst painted a 'Euclid' made – in part – out of bric-a-brac. The shuttlecock, the flowers and perhaps even the twisty cabbage-like shape of the hat all have close parallels in Arcimboldo's paintings.

But perhaps this is all too literal-minded and Ernst, like his contemporary René Magritte, tricked the viewer into seeing what he had not painted. The evidence that there is a human figure behind this assemblage of objects is rather slight. The inverted-pyramid 'head' resembles nothing so much as a mask: and masks need not have faces behind them. Ernst maintained a lifelong scepticism about received moral and aesthetic codes and their spokesmen, and about reason and rationality: empty masks and empty clothes could, for him, be signs of the vacuity of such things.

This was, relevantly, a period in which Euclid himself showed a distinct tendency to vanish from history: like certain other ancient authors. In 1966, French mathematician Jean Itard, introducing a reissue of François Peyrard's translation of the *Elements*, asserted that the book was the work not of an individual but of a group, a school, for which 'Euclid' was a collective name. (As it happens, a group of French mathematicians had been publishing under the collective name 'Bourbaki' since the 1930s, providing Itard with a relevant model for his claim about Euclid.) In the twenty-first century it is still sometimes said that the *Elements* is an accretive text for which there is no need to name an author. (This degree of scepticism does not seem entirely convincing, though it is certainly the case that the more closely one looks at the historical evidence, the more elusive Euclid becomes.) The notion of Euclid as a faceless committee – or as a figure who has deceived in the most basic way of all, by not existing – feels similar in spirit to the enigmatic, perhaps empty, mask of Ernst's non-portrait.

Euclid has a distinct context in Max Ernst's prolific output. The painting's idiom – figures with enigmatic geometrical mask-faces; quasi-geometric, patterned backgrounds; and a polished, technically perfect surface – recurred in several other works from the 1940s, including those titled *Leonardo da Vinci* and *Albertus Magnus*. These are similarly enigmatic non-portraits of Great Men associated with reason and rationality in both their dark and their light sides; there were a *Portuguese Nun* and a *Cocktail Drinker*, among several others, with similar characteristics. These highly controlled works marked what some have called a deliberate peak of technical excellence, of craft, in Ernst's output: although faceless and some-times headless – or animal-headed – apparitions had long been a motif in his work.

That the Euclidean and the non-Euclidean were on Ernst's mind in this period is confirmed by the title of *Young Man Intrigued by the Flight of a Non-Euclidean Fly* (1942–7), another painting of a mask-like triangular face, this one surrounded by the chaotic orbits of the title's 'fly', in fact produced by swinging a dripping can of paint in ellipses from a string. The technique produced results that look very like something out of an old astronomy textbook (such as the Euclidean *Phaenomena*); Ernst painted a *Bewildered Planet* in the same mode. Ernst would claim that Jackson Pollock took the drip technique from him, though not all historians have been quite convinced.

This is a compulsively ambivalent painting. Flowers or birds? Corsage or shuttlecock? Man or woman? (The hat and wrap are read more easily as feminine apparel, but Euclid is a man's name and the fish is a Freudian symbol of the masculine gender.) Foreground or background? Rational or irrational? Human figure or collection of inhuman objects?

Finally, the mask (or veil) and the woman's robe recall the old story about Euclid of Megara, long confused with the geometer, and who was a political victim like Ernst himself. Perhaps the painting is about that case of mistaken identity. Perhaps it is a painting of both Euclids at once.

Euclidean designs

A retired cartoonist covers sheets of wood with painted diagrams. A computer enthusiast recreates a nineteenth-century textbook online. A couturier is called the 'Euclid of fashion'. An Iranian exile fills sheets of paper with 'variations on the hexagon'. In the world of modernist art and design, Euclid is everywhere and nowhere.

The twentieth century saw plenty of designers, artists, theorists and critics pay tribute to the aesthetic importance of geometrical form. Lines and right angles, circles, triangles and squares turned up again and again in the forms of the modernist world. But for all that, explicit relationships with Euclid turn out to be elusive: hard to track down, and as often as not illusory.

Crockett Johnson (1906–75) provides perhaps the clearest possible case of Euclidean geometry repurposed as art. An American cartoonist and writer, he was the author of highly regarded comic strips and children's stories in the 1940s and 50s. A decade on, with the response to his books now flagging, he began a new career as an amateur painter. He was inspired in part by the architecture he had seen on a trip to Greece, and in part by *The World of Mathematics*, a 1956 collection of essays about mathematics and mathematicians. His tools were unconventional: house paint and fibreboard from a local hardware store. And all of his paintings – he would complete more than a hundred – were based on geometrical diagrams. Johnson did not call himself an artist but said instead that he 'made diagrams'.

His paintings were mostly around two or three feet square. Typically he would copy a geometrical diagram in this large size, and colour in the different parts of it with flat blocks of colour. Johnson did not put in labels for lines or points, but nevertheless many of the diagrams were easily recognisable from their geometrical sources. One of the first he completed was based on the iconic diagram accompanying Pythagoras' theorem (number 47 in Euclid's *Elements*), with three squares around a triangle. At nearly four feet square it was one of his largest paintings, and he characteristically filled in different portions of the shapes with different colours. On a dark brown background, he filled pieces of the triangle with alternating blocks of black and white, and the squares with blocks of black, yellow, blue and red. It was an eloquent, thoroughly modern testimony to the aesthetic values Crockett Johnson – like many others – had found in Euclidean geometry. He called this painting 'a kind of imposing thing . . . very pretty, too'.

Johnson made twenty-five paintings based on diagrams in *The World of Mathematics*, and later moved on to other geometry textbooks, painting the diagrams of mathematicians from ancient Greece to the twentieth century; 'a series of romantic tributes', as he put it, 'to the great geometric mathematicians from Pythagoras on up'. In his later work he stuck less closely to his sources, producing free-form fantasias of geometrical shapes, lines and

Crockett Johnson's *Proof of the Pythagorean Theorem.*

colour. But a basis in specific geometrical diagrams was always there to see. An exhibition took place in New York in April 1967 under the title 'Abstractions of Abstractions'. One art critic credited the works with 'a certain cool insouciance', adding that 'the painting is a kind of cool Hard-Edge, but bouncy overall'.

His geometrical paintings led Johnson to develop an interest in geometrical problems in their own right. He found a new geometrical approximation for the value of π, and a new way of constructing a regular heptagon, both of which were published in the *Mathematical Gazette.* His exhibitions in 1970 and 1975 displayed art based on some of this work, as well as on classic geometrical conundrums like making a square and a circle with the same area, or dividing an angle into three equal parts. Johnson died in 1975, and the bulk of his paintings are now held at the

National Museum of American History in the Smithsonian Institution; they can also be seen online.

In much the same vein – geometry as art – is the recent elevation to cult status of Oliver Byrne's edition of the *Elements*. Byrne (1810–80) was an Irish teacher of mathematics, inventor of scientific instruments and author of mathematical books. Perennially impecunious, he lived for much of his life in London; he also spent a period in New York working on a dictionary of engineering. His projects to reform mathematical teaching failed to inspire his contemporaries, and he wrote in one application for financial support that 'all of my books, inventions, and important discoveries seem only to lead me into trouble'.

One of those inventions was 'the Teaching of Geometry by Coloured Diagrams', as he put it in a pamphlet written when he was twenty-one. His plan – a contribution to the nineteenth-century debate about geometry teaching and its methods – was to replace the traditional language of lettered diagrams with coloured shapes, which he claimed would make it easier to acquire and retain geometrical knowledge. 'The use of coloured symbols, signs, and diagrams in the linear arts and science renders the process of reasoning more precise, and the attainment more expeditious.' Like many another Euclidean enthusiast, Byrne was evidently an overwhelmingly visual thinker, and for him at least, a coloured geometry book would appeal 'forcibly to the eye, the most sensitive and the most comprehensive of our external organs'.

In his pamphlet he applied his ideas to the first book of the *Elements*; by 1838 he was advertising the first six books in a coloured edition, to cost eight shillings. It was far from simple to get his coloured *Elements* into print, however, and it was nearly another decade before he did so. *The first six books of the*

Elements of Euclid in which coloured diagrams and symbols are used instead of letters for the greater ease of learners appeared in 1847, at the unfortunately exorbitant price of £1 5s. It was printed in red, yellow, blue and black, the four colours printed from four separate plates: in other words, each sheet had to be put under the press four times, and aligned precisely each time. The book failed to find buyers and the printer – one William Pickering, who had a penchant for complex, expensive projects – was declared bankrupt a few years later. Byrne was still promoting the use of colour in geometry teaching twenty years on, but in the welter of new versions of the *Elements* in Victorian Britain, his does not seem to have been much taken up by actual teachers or students.

Byrne himself denied having any aesthetic motivation: 'I do not introduce colours for the purpose of entertainment, or to amuse by certain combinations of tint and form, but to assist the mind in its researches after truth, to increase the facilities of instruction, and to diffuse permanent knowledge.' But it is as an aesthetic achievement that his book has most often been valued. It was exhibited at the Great Exhibition in London in 1851 on the strength of its beauty and the technical excellence of the printing. It continued to attract occasional mention in surveys of Victorian printing, but around 2000 it unexpectedly attained classic status. A full set of photographs of its pages appeared online, with an appreciative commentary; another enthusiast coded it into the TeX computer typesetting language (in English and Russian). The German art publisher Taschen brought out a facsimile of the book in 2010.

Late the same year Chicago-based 'designer, data geek, fractal nut' (his words) Nicholas Rougeux achieved a masterpiece of online typography, recreating Byrne's *Euclid* for the web, complete with colours, Byrne's unusual mathematical symbols, and specially designed initial letters. This time around, the shapes were not just coloured but clickable, and you could also purchase a poster showing all 269 diagrams from books 1–6 of the *Elements*. And in case that was not enough celebration of Byrne's *Euclid* as art, boutique publisher Kronecker Wallis in 2019 carried out a completion of Byrne's book, extending the coloured presentation to all thirteen books of the *Elements*.

So much for direct transformations of Euclid's *Elements* into *objets d'art*. The world of big-name modern designers and artists has also, many times, made oblique contact with Euclidean geometry: but the path from textbook to artwork has tended to be less direct and harder to follow.

The winter 1938–9 collection by fashion designer Elsa Schiaparelli (1890–1973), titled 'Zodiac', is sometimes said to have been inspired by Euclid's *Elements*. Schiaparelli was the niece of astronomer Giovanni Schiaparelli (who famously thought he saw a canal network on Mars through his telescope) and the first couturier to present complete collections with a guiding theme. From the mid-1920s she was based in Paris, and 'Zodiac' was, with 'Pagan', 'Circus' and 'Commedia dell'Arte', one of her best and most imaginative collections. *Time* magazine in 1934 placed her ahead of Coco Chanel: 'madder and more original than most of her contemporaries, Mme Schiaparelli is the one to whom the word "genius" is applied most often'.

Schiaparelli's dramatic, modern, occasionally surreal designs certainly have something geometrical to them: a piece in the *New Yorker* in 1932 spoke of 'the un-European modernity of her silhouettes, and their special applicability to a background of square-shouldered skyscrapers, of mechanics in private life and pastimes devoted to gadgetry'. The 'Zodiac' garments were emblazoned with astrological and astronomical motifs which gestured in broadly the same direction. The link with Euclid in particular seems to go back to the efforts of Schiaparelli's publicity agent, Hortense MacDonald, to describe the 'Zodiac' collection as a unified whole. Her commentary asserted that the collection was built on strict geometrical measurements, mentioning Euclid's *Elements* as a source: the slim, square-shouldered silhouettes, the slightly raised waist. Sadly it seems unclear whether the Euclidean connection was made or even assented to by Schiaparelli herself.

It is a similar story with Schiaparelli's contemporary and rival in Paris, the Italian-born Madeleine Vionnet (1876–1975). Inspired by the natural drape of the Greek *peplos*, she designed clothes whose textile shapes were often simple geometrical ones: skirts cut in full circles, a dress built from four squares (hung diagonally from the shoulders to make overlapping diamonds); another dress made of hexagons of various sizes. Her work has been compared to the Cubist break-up of picture space and, presumably on the basis of her use of simple shapes, she was and is sometimes called the 'Euclid of fashion'.

Once again Euclid lurks in the background, transformed nearly out of recognition. It seems impossible to say whether Vionnet would have avowed a Euclidean influence on her work, or thought it relevant.

One more example: equally indirect perhaps, but equally intriguing. Monir Shahroudy Farmanfarmaian (c. 1923–2019), art superstar of the Middle East, described herself as 'just a person with a good eye who happens to work with mirrors'. After her childhood in Iran, the Second World War upset her plans to study in Paris, and her training took place in Tehran and New York. The freelance commercial and fashion illustrator was inspired to take a new direction during a second period in Iran after 1957. Visiting the Shah Cheragh shrine in Shiraz, she was amazed by the mirrored hall she saw there: mirrors cut into small pieces and slivers, placed in mosaic patterns on plaster. It was a traditional Iranian craft, and Farmanfarmaian became the first modern artist to reinvent it for the twentieth century.

Farmanfarmaian collaborated with craftsmen to realise her designs, cutting mirrors into the required shapes and working up small motifs into large space-filling designs: 'I would make a drawing first, then a small sample of a motif, and then ask Haji [master craftsman Haji Ostad Mohammed Navid] to cover a certain part of the wall with it, going from one corner to the other. He would take his piece of string, dip it in charcoal dust, anchor it with his thumb like a compass, and map out the whole thing.'

Increasing success followed through the 1970s, but her possessions were confiscated during the 1979 Islamic Revolution (her husband was a descendant of shahs) and she spent the years to 2004 in the US. She worked mainly on paper, lacking the access to traditional skills to continue her mirror-and-mosaic work.

Farmanfarmaian returned to Iran in 2004, where she once again worked with craftspeople and developed her mirror mosaics into new degrees of complexity. The space-filling geometries increasingly expanded into the third dimension as relief or sculpture, and more than one commentator echoed her own view that her reinvention of geometrical shapes and their combination seemed to have infinite potential. Her work has been shown in exhibitions around the world, and in 2017 a dedicated museum opened in Tehran.

The characteristic geometrical mosaics, using small motifs to organise and fill space, are a development, a reinvention, of similar techniques with their roots in the Middle Ages, when one of the elements that went into their devising was contact between craftsmen and Euclidean geometers: such as that documented by al-Buzjani. Farmanfarmaian mentioned Iranian, Persian and particularly Sufi sources of inspiration. Of her own beginnings in this style she wrote:

> I asked my master craftsman, who had done mirror work for mosques and shrines, how he knew how to combine the motifs. He was the one who told me, long ago, that everything is geometry. He drew a hexagon and cut it into pieces and showed me how they can be combined into shapes of many sides. I had to learn about it; I went to London and bought a lot of books on geometry, and then I hired someone to teach me algebra.

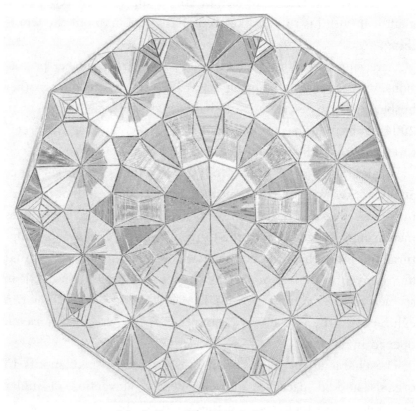

Monir Farmanfarmaian's *Decagon*.

Was Euclid's *Elements* one of the books? It would not be surprising; and even if not, it seems clear that Euclid's geometry was one of the ultimate ancestors of Farmanfarmaian's unmistakable work. Yet it was far from being the only or the most important ancestor, in a mix that included both the long Islamic tradition of geometrical design and the expertise of the craftsmen she worked with from the 1970s on. She emphasised the role of spontaneity and intuition in her own design work: 'It's not a very intellectual process.'

Lambda

Curved space, dark energy

December 2022. The Lagrange Point L2, 1.5 million kilometres from the earth. A satellite dances its slow dance, locked into synchrony – or nearly so – with the earth and the sun.

It gathers light from distant galaxies, bouncing it through an array of mirrors – three curved, three flat – then through a filter and into its two instruments. Its mission is to map the geometry of the distant universe, and study the properties of the dark energy that determines its rate of expansion.

And its name?

Euclid.

By the twentieth century, Euclid – the book, the man – had become a cultural icon, a name to commemorate and conjure with, and a shape-shifter, capable of endless reinvention and repurposing even as his specific achievements seemed to be losing their direct relevance and most of their visibility. The *Elements* was taught in ever fewer schools and universities; Euclidean geometry in any sense that Euclid would have recognised was no longer a research area for professional mathematicians. His '*quod erat demonstrandum*' and its abbreviation QED were replaced by the supposedly less aggressive box symbol (typographers call it a tombstone) for the end of a proof: ∎. But Euclid remained a name: in histories of geometry, in descriptions of physical or mental structures as Euclidean or non-Euclidean.

The *Elements* received new life as a Great Book in liberal arts programmes, and Euclidean-style geometry remained popular in books and articles for the general reader and in mathematical competitions, where its seeming universality, its accessibility, its lack of reliance on special notation or techniques, made it something of a leveller. From its start in 1959 the International Mathematical Olympiad included at least one problem in Euclidean geometry – often several – in its final paper each year.

And for historians, Euclid himself remained frustratingly elusive, his biography no more than a few lines, his existence still regularly disputed. Perhaps for that reason, there have been very few fictional accounts of his life and times, despite the occasional vogue for novelistic accounts of great scientists and thinkers. One exception was an uproarious spoof from 1922 called *Euclid's Outline of Sex*. Subtitled *A Freudian Study*, it poked robust fun at the fad for Freud-inspired rereadings of well-known figures, cooking up a Euclidean childhood dominated by the warring clans of Delta Upsilons and Phi Beta Kappas, a 'grandmother complex': in short, 'Euclid, to put it bluntly, reeks with sex.' Another was a French novel of 2002 called *Euclid's Rod* (*Le bâton d'Euclide*), which gave a version of the history of the library at Alexandria, starting with Euclid and continuing through the line of the inheritors of the stick he used in his teaching.

Euclid remained a name. A horse named Euclid ran in the 1839 St Leger Stakes; the result was a dead heat. A British cargo ship named *Euclid*, built in 1866, sank after a collision in 1890 off the coast of County Durham. Throughout the 1950s the manufacturers of construction equipment Blackwood Hodge marketed a model of tractor and scraper called Euclid, described variously as 'Gluttons for Punishment', 'Euclid the Pioneer' and 'Euclid the Empire Builder'. At their peak of popularity they were at work in more than seventy-five countries, delivering 'high speed, more loads' and 'bigger profits'.

Streets, towns, schools, journals and typefaces all bear Euclid's name. The heyday for naming things after Euclid seems to have fallen in the later nineteenth century, when any number of Euclid Avenues sprang up around North America: there are celebrated examples in Cleveland, Brooklyn and San Francisco. At certain times and places his has also been quite a popular name for people, though if there is a pattern to this its reasons are elusive. There were plenty of Euclids in – again – the nineteenth century, particularly in both North and South America; there were rather fewer in Europe, and in the twentieth century fewer still.

And *Euclid* the satellite? Edwin Hubble and others noticed in the 1920s that the light from distant galaxies was redder than it should be, as though the galaxies were receding or the intervening space was getting larger. Astronomers concluded that the universe was expanding. For most of the twentieth century it was reckoned that the relationship was linear: a galaxy twice as distant receded twice as fast. In other words, the universe was expanding at a constant rate, at least roughly.

But in 1998, studies of supernovae (certain types have a constant luminosity, and it is therefore relatively easy to be sure how far away they are) changed that view. It now seemed, with better data, that the rate of expansion was slower in the past; in other words, the expansion was accelerating. What could possibly drive the universe to expand faster and faster? The physicists' shorthand for that-which-is-driving-the-acceleration is 'dark energy'.

As well as driving the acceleration, dark energy has another job. It turns out that the structure of space on the very large scale is, perhaps unexpectedly, Euclidean. Not curved, non-Euclidean or mind-bending like Lobachevskii's or Riemann's, but flat, or very nearly so. There are wrinkles wherever there is mass, but the overall picture is Euclidean. To be flat, the universe must contain a certain amount of mass: holding it flat, so to speak. But the matter that can be seen, and the dark matter that can be inferred from grav-itational observations of galaxies, accounts for only a little more than a quarter of the mass needed. The remainder – nearly three-quarters of the mass in the universe – is dark energy.

For dark energy to be more than a fudge factor it needs to have detailed properties, which need observations. It permeates space, but at very low average density; it exerts a gravitational effect but it is unlikely to be detectable in the lab; you can think of it as the 'mass' of empty space. Is it a constant presence in all empty space at all times? Or does the observed acceleration of the universe's expansion require a more complex model, a dark energy that varies across space and time? In the first case, dark energy would amount to a minimal modification of the accepted understanding of gravity: the variable Λ (lambda) that Einstein once added to his equations and then deleted, reckoning it an unnecessary mistake. If the second possibility is right, though, then either fundamental physics or general relativity (or both) might need more serious modification to accommodate the new, omnipresent field of dark energy.

Enter *Euclid* the satellite, whose purpose will be to help answer these and related questions. In 2007 the European Space Agency issued a call for proposed missions, and two of the responses addressed the geometry of the distant universe: the 'Dark Universe Explorer' and the 'Spectroscopic All Sky Cosmic Explorer'. *Euclid* was the combined result, adopted by the ESA in 2011–12 and developed through several years of specification and review. Hardware began to be delivered in 2017, but a problem with the electronics behind one of the detectors delayed the launch for a few years. Meanwhile a consortium involving labs and agencies in fifteen countries has been assembled to manage the project and plan ways to deal with the enormous amount of data the mission will generate.

If everything goes to plan, the *Euclid* spacecraft, weighing a couple of tons, will launch from the ESA's space centre in French Guiana in June 2022. After about thirty days of travel it will stabilise around the L2 Lagrange point, orbiting the sun at the same rate as the earth but a million and a half miles further away. There it will remain for the six years of its mission, with a clear view of deep space away from the light, heat, magnetic and gravitational disturbances of the earth and the moon. A four-foot curved mirror – one of the most accurate ever made – will collect the light from distant galaxies and send it to a pair of instruments: a visible-light camera and a near-infrared one. The pictures will come back to the earth as radio signals: around a hundred gigabytes of data every day.

The high-resolution pictures in visible light will be used to measure how the images of a billion distant galaxies are distorted by the mass of objects in front of them, enabling scientists to build a new map of mass distribution in the universe. The infrared images will be used to study the spectra of tens of millions of distant galaxies, and to determine their speeds and distances, improving on the 1998 results for the history of the cosmic expansion. A million of the spectra will be measured with higher accuracy in concert with ground-based observations, in order to study the typical large-scale clustering of distant galaxies, structures which are a frozen record of density waves in the primordial plasma that preceded galaxy formation. All of which will help to understand the – possibly changing – properties and effects of dark energy in the expanding universe, and the geometry – Euclidean or not – of space on the largest scale.

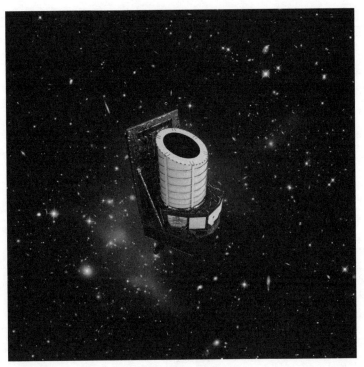

Euclid, the satellite.

Epilogue

The story of Euclid's *Elements* is about what ideas can be and what they can do, and how far a book can travel, evolve and thrive. It is a story about mathematics and its place in the lives and minds of people in many cultures. It is the story of a North African geometry textbook that has impacted on the world for 2,300 years.

That book has been on a fantastic journey. It has found readers around the world who have encountered, engaged and done things with and to it that were meaningful to them. Over time, the very austerity of the text has enabled people to find in it whatever they needed. For some, Euclid was the great author of a Great Book: a classic work of Greek literature. For others he was a philosopher whose work could guide readers into the mysteries of knowledge and existence. For others again he was a hero of practical life who could teach how to design a cathedral or paint a picture.

Under the pressures of the Enlightenment world Euclid – like other ancient authors and authorities – came under suspicion, his traditional virtues collapsing but later recovering, as he came to be recognised as a shape-shifter capable of almost anything, his very malleability now one of his virtues. If the twentieth century (and, so far, the twenty-first) has seen his book somewhat eclipsed – suspected, distrusted, disused – yet it has also seen a new interest in embracing its ambiguities and its uncertainties: playing with it, and allowing it to play with its readers.

No one can see what will happen to Euclid's *Elements* next, as cultures change and change again. But the book seems adaptable like no other, with more ideas to it than will fit behind any mask: ideally suited, perhaps, to the fluid, fragmented world of the information age. It is easy to imagine it dancing on into the future, never fully grasped by any one of the cultures it touches. Surely it will go on finding more lives to live: ancient and new, wise and useful, tricky, playful, and always just slightly out of reach.

Acknowledgements

It is a pleasure to acknowledge the help of those who contributed to the long process that brought this book from idea to realisation. Victoria Kwee and Caroline Davidson contributed many hours and many ideas to the drafting of early versions of the book proposal. Years later Arabella Pike believed in the book on the slenderest of evidence, and Felicity Bryan shepherded it through the commissioning process. My thanks to Arabella and to all of the team at William Collins who worked on the book.

Special thanks are due to Yelda Nasifoglu and Philip Beeley, who worked with me for two years on 'Euclid in Early Modern Britain'; some of the research done during that project, and some of their ideas, have found their way into this book. The participants at our two workshops in 2016 and 2017 did more than they realised to inform and shape my thinking about Euclid and his *Elements*, as did those at the History of Mathematics Forum and the seminar on the History of the Exact Sciences at Oxford during its long gestation. I am particularly grateful to Richard Lawrence at the Oxford Bibliographical Press for the opportunity to set and print a page of the *Elements* by hand.

My thanks also to the Arts and Humanities Research Council for supporting our research project, and to All Souls College for its support of my work over many years. I am grateful to the staff at the Bodleian Library and particularly its Rare Books and Manuscripts Reading Room for help with a variety of sometimes obscure requests.

Anna-Marie Roos and Chris Hollings were kind enough to read the book in typescript, saving me from many blunders and improving it in numerous ways. The book's remaining defects are of course my responsibility alone.

Among my family, Jessica, William, Ralph and Laurence kept me sane, and my parents as ever read drafts with enthusiasm and understanding.

Image Credits

Page 8. Silver stater of Ptolemy I, 305–285 BC. Cleveland Museum of Art 1916.994. *(Creative Commons/Public Domain, CC0 1.0)*

Page 13. Christoph Clavius, *Euclidis elementorum libri* XV (Rome, 1574), fol. 21v. *(Author's collection. Image © Benjamin Wardhaugh)*

Page 21. Ostrakon, Elephantine, third century BC. Berliner Papyrusdatenbank P. 11999. *(© bpk-Bildagentur)*

Page 28. Luca Pacioli, *Divina proportione* (Venice, 1509), plates XXII and XXVIII. Getty Research Institute, 47289. *(Internet Archive/Public Domain)*

Page 44. Oxford, Bodleian Library, MS D'Orville 301, fol. 268r. *(The Picture Art Collection / Alamy Stock Photo)*

Page 50. Francis Rawdon Chesney, *Narrative of the Euphrates Expedition carried on by Order of the British Government during the years 1835, 1836, and 1837* (London, 1868), p. 406. *(Archivac / Alamy Stock Photo)*

Page 58. Leiden University Libraries, SCA 1, fol. 2r. *(Leiden University Libraries/Creative Commons Attribution International, CC-BY 4.0)*

Page 66. Werner Rolevinck, *Fasciculum temporum* (Venice, 1485), fol. 37v. *(Digital Library@Villanova University, Creative Commons Attribution-ShareAlike 4.0 International, CC BY-SA 4.0)*

Page 71. *Preclarissimus liber elementorum Euclidis perspicacissimi* (Venice, 1482), fol. 2r. *(Wikimedia Commons/public domain)*

Page 88. *Euclidis quae supersunt omnia* (Oxford, 1703), title page. *(BEIC digital library/public domain)*

Page 98. Roman mosaic of the first century BC, from the house of T. Siminius Stephanus, in Pompeii, now at the Museo Nazionale Archeologico, Naples. *(Jebulon, Creative Commons/Public Domain, CC0 1.0)*

Page 110. Johann Georg Leuckfeld, *Antiquitates Gandersheimenses* (Wolfenbüttel, 1709), plate facing 271. *(Wikimedia Commons/public domain)*

Page 124. Francesco Villamena, portrait engraving of Christoph Clavius, 1606. *(Rijksmuseum Amsterdam, Creative Commons/Public Domain,CC0 1.0)*

Page 134. Anonymous portrait of Xu Guangqi at Guangqi Park, Shanghai. *(Mountain, Creative Commons Attribution-Share Alike 3.0)*

Page 154. Unattributed portrait of Baruch Spinoza, c.1665. *(Wikimedia Commons/public domain)*

Page 166. Joshua Horner, portrait of Anne Lister, c.1830. *(Wikimedia Commons/public domain)*

Page 175. Wall painting from the Tomb of Menna at Thebes, c.1400 BC *(www.BibleLandPictures.com / Alamy Stock Photo)*

Page 182. Terracotta amphora attributed to the Berlin Painter, c.490 BC. The Metropolitan Museum of Art, 56.171.38. *(Metropolitan Museum of Art/Public Domain, CC0 1.0)*

Page 190. Herzog August Bibliothek Wolfenbüttel, Cod. Guelf. 36.23 Aug. 2°, fol. 66v. *(HAB Wolfenbüttel/Creative Commons, CC BY-SA)*

Page 200. Tiled panel in the Jemah Mosque, Isfahan. *(© Alamy Stock Photo)*

Page 202. Statues of Geometria and Euclid, Chartres Cathedral: West façade, South Arch, c.1150. *(By kind permission of Nick Thompson)*

Page 206. British Library, MS Burney 275, fol. 293r. *(British Library/public domain)*

Page 208. Brussels, Bibliothèque royale, MS IV, 111, fol. 90v. (© KBR)

Page 210. Bodleian Library, MS Ashmole 304, fol. 2r. *(Bodleian Libraries/ public domain)*

Page 213. Piero della Francesca: *Flagellation of Christ. (Wikimedia Commons/public domain)*

Page 232. Trinity College, Cambridge, NQ.16.201[1]: Isaac Barrow, *Euclidis Elementorum libri XV* (Cambridge, 1655), p. 146. *(By permission of the Master and Fellows of Trinity College Cambridge)*

Page 247. Thomas Phillips, portrait of Mary Fairfax, Mrs William Somerville. *(Wikimedia Commons/public domain)*

Page 270. *The Works of George Eliot* (New York, 1910), vol. 3: *The Mill on the Floss*, frontispiece. *(Wikimedia Commons/public domain)*

Page 279. *Taḥrīr-i Uqlīdis* (Mathurā, 1884), p. 20. *(Bodleian Libraries/ Google Books, CC BY-NC-SA 2.0)*

Page 292. Elliott & Fry, portrait of Sir Thomas Little Heath. National Portrait Gallery x89686. *(© National Portrait Gallery, London)*

Page 301. Max Ernst, *Euclid* (1945). *(Harry Croner/ullstein bild via Getty Images)*

Page 308. Crockett Johnson, *Proof of the Pythagorean Theorem.* Smithsonian Institution 1979.1093.01. *(Division of Medicine and Science, National Museum of American History, Smithsonian Institution)*

Page 314. Monir Shahroudy Farmanfarmaian, Decagon (Third Family), 2011, Mirror and reversed glass painting on plaster and wood, 120 cm diameter. *(Courtesy of the Third Line)*

Page 321. Artist's impression of the Euclid spacecraft. *(Image © ESA/ ATG medialab (spacecraft); NASA, ESA, CXC, C. Ma, H. Ebeling and E. Barrett (University of Hawaii/IfA), et al. and STScI (background))*

Notes on Sources

This is a selective indication of my main sources of information for each chapter, including the sources of direct quotations where these are not clearly indicated in the text. Minor or supplementary sources are not listed exhaustively, and standard editions of the works of well-known authors are not normally mentioned.

Abbreviations

BNP Cancik, Hubert, et al. (eds), *Brill's New Pauly* (Leiden, 2012).

DSB Gillispie, Charles Coulston, et al. (eds), *Complete Dictionary of Scientific Biography* (Detroit, 2008).

EE Euclid's *Elements*.

OCD Hornblower, Simon and Antony Spawforth, *Oxford Classical Dictionary* (3rd edition, Oxford, 2005).

Alexandria

The encounter of Euclid and Ptolemy appears in Proclus's *Commentary on Elements I*, 55–6; the version involving Menaechmus and Alexander, as well as the 'threepence' story, are in Stobaeus, *Eclogae* II.

Ptolemy is discussed in Ellis 1994 as well as Hölbl 2001 and Green 2008. On Alexandria I have referred among other works to Marlowe 1971, Fraser 1972, Haas 1997, Erskine 2003 (p. 483 for 'wealth, wrestling schools', from the first mimiamb of Herodas) and Hirst and Silk 2004. On the Library and Museum see also El-Abbadi 1990, Blum 1991 and MacLeod 2002 (p. 62 for 'well-fed bookworms', from Athenaeus, *Deipnosophistae* 22d). The lines from Cicero are from *De natura deorum* 2 and *Tusculan Disputations* 5.23.

On Greek science see Tupin and Rigll 2002 as well as Lloyd 1973. Greek mathematics including *EE* and its composition are discussed in Cuomo 2001, also in works by Netz (such as 1997, 1999, 2002, 2004); on geometry as a game see also Asper 2009. 'How to draw an equilateral triangle' is a loose paraphrase of *Elements* 1.1; 'evident even to an ass' from Proclus, *Commentary* (322 tr. Morrow).

The best succinct discussion of Euclid's life and dates is in Fraser 1972 (vol. 2, p. 563); Heath 1926 also gathers the surviving testimonies and later anecdotes. *DSB*, *OCD* and *BNP* take various views on the degree of certainty attaching to the extremely scanty biographical data. A lucid account of Euclid's other works appears in the introduction (by Maurice Caveing) to Vitrac 1990.

Elephantine

Papyrus, its characteristics and durability as well as the issues of the interpretation of fragments are discussed in Černy 1952, Turner 1968, Lewis 1974 (the story of the Berlin curator is at p. 58) and Erskine 2003; further details come from Roberts and Skeat 1983, Turner 1987, Netz 2002b, Rowlandson and Harker 2004 and Meskens 2010.

Ostraka (and ostracism) are the subject of helpful articles in *BNP* and *OCD*; Fowler 1999 has a handy list of the early fragments of *EE* (pp. 210–16), as does Netz 1998 (p. 274); see also Stamates 1969 (vol. 1, pp.

187–8). More general issues in the early spread of texts and their audiences are addressed in Lemerle 1986 and *BNP* (s.v. *book*), and with reference to mathematics in Netz 1997 and Cuomo 2001.

The ostraka I discuss from Elephantine (P. 11999 in the Berliner Papyrusdatenbank) are published in Mau and Muller 1960; their specifics are addressed as well in Fraser 1972 (v. 2, p. 558, n. 43), Fowler 1999 and Cuomo 2001 (p. 72 for the quotation, see also p. 145). Background information on Elephantine is in *BNP* and *OCD* as well as Porten 1996; I have also used Lewis 1986 (pp. 21–4) on Hellenistic garrisons.

Hypsicles

The confiscation of books by Ptolemy Euergetes is described in Galen's *Commentarium in Hippocratis Epidemiai* 3.239–40.

For the development of Greek mathematics during this period I follow in the main the interpretations of Netz (particularly 2008); see also Cuomo 2001. As well as the works of the mathematicians mentioned in the text, for specific details see Pfeiffer 1968 (pp. 152–6), *DSB* (s.v. *Euclid*), Bulmer-Thomas 1973, Lloyd 1973, Macleod 2002 (p. 6), Acerbi 2003 and Horowitz 2005 (s.v. *Geometry*).

For Hypsicles himself, aside from *OCD*, *DSB* and *BNP*, all other discussions are now superseded by Vitrac and Djebbar 2011, whose account of the composition of what is now *Elements* XIV I follow.

Theon of Alexandria

For the evolution of Alexandria see Marlowe 1971, Bagnall 1995 and Haas 1997, with some information from Ellis 1994 and Green 2008. On the Library and Museum in particular see Butler 1978, El-Abbadi 1990, Blum 1991, Cameron 1993, MacLeod 2002 and Hirst and Silk 2004. The physicality of scribal practice in this period is discussed in Turner 1987 (pp. 4–7). These sources (and *OCD* and *BNP*) take various views on the episodes of destruction from the second century BC to the fourth century AD (the Brucheion as 'desert' is from Epiphanius; see Lloyd-Jones 1990, p. 29); on the absurd fable blaming the destruction of the library on the Arab conquest in the seventh century there is no need to look beyond the thorough discussion in Butler 1978 (pp. 401–26).

The philological tradition of this period is addressed in Pfeiffer 1968 and (particularly) Blum 1991, with Reynolds and Wilson 1968/2013 and Olmos 2012. The Euclidean commentary tradition (on which see also the chapter on Proclus in this book) is discussed in Heath 1926, Knorr 1989, 1996, Russo 1998, Cuomo 2001, Vitrac 2001, 2004, Acerbi 2003, Netz 2012, Keyser and Irby 2008 (s.v. *Euclid*) and Chemla 2012 (prologue). These sources, particularly those by Knorr and Vitrac, also discuss what can be known about Theon's edition of *EE*. As Vitrac (2006) says with more specific reference to the Arabic evidence, 'the time for synthesis and certainty has not yet arrived'.

For Theon himself, in addition to *BNP*, *DSB*, Keyser and Irby 2008 and *OCD* (whence 'trivial reworkings' and 'completely unoriginal'), see Wilson 1983 (p. 42), Cameron 1993 and Dzielska 1995; the latter is also by far the most useful work on the perennially misrepresented Hypatia, some discussion of whose work also appears in Knorr 1989. 'We have proved' is from Theon's commentary on the *Almagest*, book 1, chapter 10; 'mathematically banal' from Muller in Pépin and Saffrey 1987.

Stephanos the scribe

On the Byzantine world and transmission and the curriculum therein see Irigoin 1962, Reynolds and Wilson 1968/2013, Wilson 1983, Lemerle 1986, Haas 1997 and the essays by Ward and Lieu in MacLeod 2002. The image of the world collapsing into a single city is from Netz 1997 (p. 15). On the seven liberal arts see also Pfeiffer 1968 (p. 52) and on *EE* therein see *BNP* (supplement, s.v. *Euclid*).

On the transmission of *EE* in this period see *DSB* (s.v. *Euclid: Transmission of the Elements*), Heath 1926 (the incorrect scribal expansion is mentioned at vol. 2, p. 109) and Acerbi 2003; I mainly follow the views of Vitrac (2010, 2012; also 2003, 2004, 2006) on the substitution and (generally later) accumulation of alternatives in the Euclidean text, as well as the relationship of the latter process to that of transliteration: but see further the chapter below on François Peyrard for more on the continuing uncertainties of the text's history. The spurious book XV is discussed in Vitrac and Djebbar 2012 and in Heath 1926.

As well as Irigoin 1962, Reynolds and Wilson 1968/2013, Wilson 1983 and Lemerle 1986, Arethas is also discussed in the introduction to Westerink 1968. The quotation about his Aurelius source is in Montanari *et al.* 2014 (p. 343), which also has a useful discussion of his scholia.

Stephanos is discussed in Aletta 2004, as is his Euclid manuscript; other descriptions of the manuscript, with identification of the various annotating hands and transcriptions of some of the notes, are in Madan 1897 (vol. 4, p. 104), Heath 1926 (vol. 1, p. 47), Weitzmann 1947 (p. 120 with plate 107), Wilson 1983 (p. 121) and Lemerle 1986 (p. 260; see also p. 197). The intriguing Arabic numerals in these annotations are discussed by Wilson (pp. 400–4). Murdoch 1966 contains a suggestion about the manuscript's later history which has not been widely taken up. Hunt 1975 discusses the history of the D'Orville collection (see pp. vii and 35), as does Vioque 2017. On D'Orville see futher van der Aa 1852 and Madan 1897 (vol. 4, p. 37). The celebrated Euclid manuscript is also mentioned in a total of four letters in Early Modern Letters Online (emlo. bodleian.ox.ac.uk), from one of which (Antonio Francesco Gori to Jacques Philippe D'Orville, 5 November 1748) I derive its probable date of acquisition.

al-Hajjaj

On the history of Islam in this period I have found Lassner 1980, Kennedy 1981 and Robinson 2010 helpful; on Baghdad in particular Lassner 1970 (p. 108 for 'in the entire world', quoting Al-Khatib) and Kennedy 2005 (p. 200 for 'I saw the caliph', the words of grammarian and literary critic Tha'lab). The translation movement is studied in more depth in Rosenthal

1975, Endress 1997, Gutas 1998 (p. 32 for 'they had heard some mention', quoting Ibn-Haldun), Abattouy 2001 and Giliot 2012.

The transmission of *EE* to the Islamic world is the subject of a substantial literature including many works by Djebbar (such as 1996, 2003), De Young (1984, 1992, 2002/3, 2004, 2008a, 2008b) and Brentjes (for instance 1993, 2006, 2008). See also Lo Bello 2003, Berggren and Van Brummelen 2005 and Elior 2018, and on Islamic mathematics and Euclidean geometry more generally Knorr 1983, Berggren 1986, Hogendijk 1993 and De Young 2002, 2009. Manuscripts are listed in Lo Bello 2003.

The names of Euclidean propositions and their origins are discussed in Kunitzsch 1993, superseding the discussion in Heath 1926.

Adelard

The literature on Euclid in Latin is now a large one, but the basic studies of Haskins (1924, 1927) on the twelfth-century translators are still of value (see Alverny 1994, chapter II for some revisions). The studies collected in Folkerts 2003 and 2006 summarise much of what is now known, while the definitive word on many questions is to be found in the critical editions of Busard (1977, 1983, 1984, 1992, 1996, 2001, 2005). Busard 2005 (pp. 1–40) seems to be the best summary of what is now known about the Latin Euclids.

On Roman bilingualism Marrou 1948 is useful; also Lemerle 1986 and Reynolds and Wilson 1968/2013 on its decline. On Adelard himself, Burnett 1987 is much the most valuable resource ('nothing in his life' is from Gibson therein, p. 13); Haskins 1927 (pp. 20–42) is still worth consulting; see also Burnett 1993, Alverny 1994 (ch. II) and Mantas-España 2014.

Erhard Ratdolt

On the practice of early modern print I am indebted to Richard Lawrence at the Bibliographic Press in Oxford for hands-on demonstrations.

On Ratdolt's life the classic study Redgrave 1894 is still important; see also Diehl 1933, Nordqvist 1961, (esp. 40–7 on the Euclid) and Risk 1982 (p. 15 for 'I was conducted', quoting Philip de Comines; p. 34 for 'the seven arts', quoting J.L. Santritter). On the impact of print more generally I have used Febvre and Martin 1958/1976, Pfeiffer 1976, Reynolds and Wilson 1968/2013, Chartier 1985, Grafton 1997 and Raven 2015. The printing of Ratdolt's Euclid is discussed in detail in Baldasso 2009 and 2013, although Baldasso's view about the printing of the diagrams has not yet found universal acceptance; see Nasifoglu (2020) for further discussion.

On early editions of Euclid more generally see Thomas-Stanford 1926, Steck 1981 and Wardhaugh 2020b.

Marget Seymer her hand

On Euclidean annotations I have used the data gathered by the 'Reading Euclid' project (www.readingeuclid.org); I am grateful to Yelda Nasifoglu for finding and reporting Marget Seymer's signature in the National Library of Wales (shelfmark: Euclid 1543). 'Manifolde additions' is from the title page of Billingsley 1570. Relevant publications on mathematical and Euclidean annotation include Wardhaugh 2015, 2019a, 2019b as well as the other essays in Beeley, Nasifoglu and Wardhaugh 2020. On the more general phenomena of early modern annotation, Jackson 2001 is an indispensible guide. Sandie Hume's signature appears in the Radcliffe Science Library, shelfmark 1831 e.4.

Edward Bernard

The basic source for the life of Bernard is Smith 1704; the article in Matthew and Harrison 2004 is useful. The only substantial discussion of his Euclid is Beeley 2020, on which this chapter is based together with an examination of the several copies of the *Elements* owned and annotated by Bernard and his circle now housed in the Bodleian library. 'Degenrous age' and 'lay aside all thoughts' are from Thomas Smith's letters to Edward Bernard dated 13/11/1694 and 30/4/1695 in Oxford, Bodleian Library, MS Smith 57; 'take care of the Geometry' from a 1698 memorandum of Henry Aldrich and others in MS Ballard 49, fol. 228r. The 'polyglot' volume described in the text has shelfmark Auct. S.1.14,15.

Gregory's mathematical work is discussed in Eagles 1977 and the preface to his 1703 edition of the works of Euclid is a useful source. More general discussions of textual scholarship in this period are Pfeiffer 1976, Reynolds and Wilson 1968/2013, Grafton 1991, 1997, Eisenstein 2005 and Maclean 2012. The 'perfect body' is from Savile 1621, p. 140, on which see Goulding 2010.

Plato

On geometry and its cultural location in ancient Greece see Netz 1999, 2004, 2008 and in particular 2012; he discusses the passage from *Meno* 82ff in (2003) (I use the translation of Lamb here, while 'if the truth about reality' is from *Meno* 86b tr. Guthrie). Briefer but exceptionally helpful is Cuomo 2001; the older accounts of Heath (1921 and 1949) must now be used with some caution. Other reconstructions of mathematics before Euclid – sometimes widely divergent – are Sachs 1917, Becker 1936, Szabó 1978, Neuenschwander 1973, 1974, Knorr 1975, Artmann 1991 and Fowler 1999. More recent scholarship on the subject is represented in Unguru 1975, Zhmud 2002 and Corry 2013.

The story about Euclid of Megara is from Aulus Gellius, *Noctes Atticae* 7.10 in the translation of Rolfe. On the philosophisation of Euclid I have used in particular Zhmud 1998 and 2002. 'Let no one come to our schools' is from al-Qifti, *Ta'rikh al-Hukama*, quoted in Heath 1926, vol. 1, p. 4; 'belonged to the persuasion of Plato' from Proclus, *Commentary* tr. Morrow, 68.

Proclus Diadochus

The main source for the life of Proclus is the biography/hagiography by his successor Marinus, translated in Edwards 2000 and in Rosán 1949 (11–32); see also Blumenthal 1984 for the interpretation of the text. Frantz, Thompson and Traulos 1988 (42–4) describes the archaeological site which may be Proclus' house, although the identification has not found universal acceptance. The section from the hymn to Minerva is from the translation of Taylor (1793).

On Proclus' philosophy, including his sources, see Siorvanes 1996 and particularly O'Meara 1989; also the essays in Pépin and Saffrey 1987, especially that by Mueller, and Claessens 2012. On Neoplatonism more generally, Wallis 1995 is exceptionally helpful; see also Glucker 1978.

On Proclus' commentary, Mueller's introduction to the translation of Morrow is helpful; so are Nikulin 2008, Harari 2006 and Ver Eecke 1948; on its sources also Mueller 1987. On the division of the mathematical proposition I follow Netz 1999. The historical material in the commentary is discussed in Cuomo 2001 (p. 260), and Zhmud 2002. Wider discussions of the logical structure of *EE* include Grattan-Guinness 1996 and the indispensible Mueller 1981.

For the *ageometretos* inscription see Saffrey 1968, and more generally Netz 2003 on the mathematization of Plato. For a modern discussion of Euclid of Megara see the *BNP*; the confusion with Euclid the geometer is also discussed in Goulding 2010.

Hroswitha of Gandersheim

Various portions of Hroswitha's work are translated in St John 1923 (p. xix for the 'ray of Sophocles'), Wiegand 1936, Wilson 1998, and Bonfante and Chipok 2013; the passages from *Sapientia* are my own translation.

The transmission of this portion of Greek number theory is discussed in D'Ooge, Robbins and Karpinski 1926 and Masi 1983 as well as Heath 1926 (volume 2).

Discussions of Hroswitha's life and works include Coulter 1929, Fife 1947, Burgess 1957, Haight 1965 (p. 23 for the 'perspicacious mind'), Frankforter 1979 and Wilson 1984. More recent readings, particularly of the martyrdom plays, include Carlson 1998 and McDonald-Miranda 2010. On particular points including her humour and her sources I have also consulted Wedeck 1928 and Coffman 1931.

Rabbi Levi ben Gershom

On ben Gershom's life and his cultural background see Goldstein 1978, Freudenthal 1995 and Feldman 2010; also Dahan 1991, certain of the essays in Freudenthal 2005 (p. 741 for 'many great and noble Christians') and Glasner 2015. On his work and philosophy Kellner 1994, Mancha 1997, Freudenthal 1992 and Touati 1973. On his mathematics, Carlebach 1910 and Simonson 2000. Kellner 1995 and ben Gershom 1998 are particularly helpful for his philosophy of mathematics. 'It has seemed good' is the opening of his commentary on *Elements* 1–5 (translated from Lévy 1992, p. 95); 'that a straight line can be extended' is slightly modified from the version in Katz 2016, p. 327.

Mathematics in Hebrew culture is discussed in Lévy 1997a, 2011 and Katz 2016; the portion of ben Gershom's Euclidean work included in the last may be read alongside the French version and commentary in Lévy 1992a, and the discussion in Lévy 1992b. On Euclid in Hebrew more generally see Elior 2018 and Lévy 1997b, 1997c.

On the later development of the parallel postulate see the chapter on Lobachevskii below.

Christoph Clavius

On Clavius and his work I have used Lattis 1994 in particular ('a man untiring in his studies' is from p. 22, as is the 'living quarters' anecdote; 'the Euclid of his times' from p. 3 and 'immortal fame' from p. 7), as well as Feingold 2002 which also discusses the context and significance of Jesuit science; on the latter subject Wallace 1984 is an important account.

For Clavius's Euclid, the reading by Price (2017) is indispensable, as is that of Rommevaux (2005). On points of detail I have consulted Knobloch 1988, Jardine 1979 and Smolarski 2002. 'Nobody can accede to metaphysics' is from Engelfriet 1993, 115. The copy I describe is in a private collection (mine); the one with the 'wonders of God' inscription (in Latin) in Caius College, Cambridge, with shelfmark M.25.19.

From the increasingly large literature on early modern editions of the *Elements* I have found de Risi (2016) most important (the examples of new axioms are quoted from here); there are summaries in Steck 1981 and Wardhaugh 2020b.

Xu Guangqi

On Xu Guangqi and Ricci respectively Jami, Engelfriet and Blue 2001 and R.P. Hsia 2010 are invaluable (from the former, p. 34, 'could speak of nothing else'; p. 383, 'no need to doubt … one cannot elude it'; from the latter, p. 248, 'the only gentleman in the world'; p. 156, 'Man of the Mountain'; p. 269, 'Man of Paradox'; p. 199, 'like madmen'). Spence 1984 is also valuable on Ricci (p. 115 for robes 'of purple silk'), as is the detailed entry in Goodrich and Fang 1976. On the Jesuit mission in China more generally I have found F. Hsia 2010 very helpful; also Gernet 1985, and Zhang 2015 on scientific exchange (p. 43 for 'besieged'). For a different account see Hart 2013 (and the reviews Hsia 2015 and Bréard 2016).

On the Xu/Ricci Euclid, Engelfriet 1998 is the standard and exhaustive account; quotations from the prefaces to the *Elements* are from pp. 292, 457–9, and the longer passage from Engelfriet himself from p. 206. I have also used Engelfriet 1993 (p. 114 for 'at my university'), Martzloff 1995 and D'Elia 1960. For the more general context of Euclidean/European geometry in this context see Martzloff 1981, Chemla 1996 and Jami 1996, 2011.

Blame not our author

'Blame not our author' is published in Gossett 1983, whence all quotations from the text are taken; it is discussed in the preface to that edition, in Mazzio 2004 (p. 49: 'anything but normal') and Nasifoglu 2017. In my quotations I modernise some spelling and punctuation.

Baruch Spinoza

Among the large number of general works on Spinoza I have found the following helpful: Garrett 1996, Nadler 1999, Viljanen 2011 and Della Rocca 2018. On the *Ethics* in particular I have used Lloyd 1996 ('Cursed be he' appears on p. 1, quoting the 27 July 1656 excommunication), Koistinen 2009 and Lord 2010. The geometrical/axiomatic method is the subject of a large literature and the question of its relationship with the contents of philosophy in any given case is frequently a disputed one; a bibliography appears in Steenbakkers 2009 (p. 53 n. 6), which itself is a helpful summary of the Spinozan case. Mark 1975 is a useful earlier summary; see also Solère 2003.

I have on the whole followed the line of interpretation associated with Gueroult (as against that of Wolfson), and have found Nadler 2006 and Byrne 2007 particularly helpful; see also for instance McKeon 1930, De Dijn 1986 and Garrett 2003. I have also used in particular (with some reservations) the reading of Curley 1988.

The story about Hobbes is quoted from Hobbes 1660 (and see also Jesseph 2004, 193), and his remarks on the advantages of geometry from the preface to Hobbes 1647; see also Rocca 2018, p. 30. Berkeley's comment is from Berkeley 1734, §2, Descartes' 'engage the minds' and 'however argumentative or stubborn' are from the *Meditations* (the second set of replies) in the translation of Cottingham, Stoothoff and Murdoch. 'That of the Mathematicians', from Meyer's preface to Spinoza's Cartesian text, is quoted in Rocca 2018, p. 21. 'Knowledge of an effect …' is Axiom IV of the *Ethics*; 'things could have been produced' is book I, proposition 33;

'from God's supreme power' from the note to book I, proposition 17; 'the order and connection of ideas' is book II, proposition 7. 'Hocus-pocus', from *Jenseits von Gut und Böse* 1.5, is quoted in Koistinen 2009, p. 43.

Anne Lister

For the general phenomenon of mathematics as a meditative and self-improving practice Dear 1995 and Jones 2006 are invaluable. Isham is studied and edited in Isham 1971 (somewhat less satisfactorily in Isham 1875), the Conways in Nicolson and Hutton 1992 (quotations are from pp. 146, 231). The line from Locke appears in Locke 1706, p. 30; the story from Vitruvius in *De architectura* 6.

On the *Ladies' Diary* the key sources are Costa 2000, 2002a and 2002b; Perl 1979 and Albree and Brown 2009 are also of value, as are the reprints of Hutton 1775 and Leybourne 1817, relatively more accessible than the original diaries.

Excerpts from Anne Lister's diaries are printed in Lister 1988, 1992, 2010 and Liddington 1998; selected letters in Lister 1992. The studies Eisner 2001 (p. 31: 'unlike a gentleman'), Liddington 1993, 1994, 1996, Hughes 2014 are useful, as is the biography Steidele 2018. I am grateful to Caroline Davidson for alerting me to Lister's reading of Euclid, and to Helena Whitbread for more detailed information including the date when Anne 'began Euclid' and the 'very stupid' quotation. Her later intention to 'proceed diligently' and 'turn my attention', and the proposed division of her day, are quoted from the entry for 13 May 1817, in Lister 1988, p. 6.

Petechonsis

The survey of Petechonsis' land is described in Fowler 1999 (pp. 231–4, based on O. Bodl. ii 1847; quotation p. 231); further sources are transcribed, translated and discussed in Thompson/Crawford 1971 and Cuvigny 1985. 'Diverse weights' is Proverbs 20:10 (NRSV). 'Divided the country' is from Herodotus II.109, in the translation of Godley; see Imhausen 2016, pp. 117–18, whence also (p. 159) the quotation from the Teaching of Amenemope, chapter 6 ('Do not move the markers').

The question of transmission is further discussed in Robson 2005; on the origins and nature of the Greek deductive tradition see Netz 1999 and Cuomo 2001; for the two cultures of mathematics in Greece see Netz 2002 and Asper 2009.

Dividing the monochord

The best survey of Euclid's works other than the *Elements* appears in the introduction to Vitrac 1990. On the *Phaenomena* and the *De ponderibus* see the monographs Berggren 2006 and Moody and Clagett 1952; on the *Optics* Burton 1945 provides an English translation, while Webster 2014 and Tobin 1990 are useful discussions.

Ancient Greek music in general, including its cultural context, are introduced in West 1992 (p. 373: 'a rhapsode ...', and also information about the career of Nicocles) and Barker 1984, 1989; for details about the *Sectio canonis* see also Barker 1981 and 2007 with Barbera 1991.

Hyginus

The *agrimensores* tradition is discussed quite fully in Cuomo 2001; a full-length treatment is Dilke 1971 (p. 45 for the passage from Cassiodorus). See also Geymonat 2009 and Lewis 2001. The texts of the *Corpus agrimensorum* are edited in Rudorff, Lachmann and Blume 1848 and translated in Campbell 2000, which has a helpful introduction and commentary. On MS Arcerianus A the facsimile and introduction Butzmann 1970 are invaluable; see also in particular Carder 1978.

The later evolution of Latin geometry teaching is addressed in Evans 1977 and Zenner 2002; for the theological side I follow Zaitsev 1999 (p. 530 quotes 'Euclid's intention is twofold' and p. 531 'approach the heavens', both from *Geometria I*). A different interpretation of the evidence is given by Stevens (2002, 2004), although his critical edition promised in 2004 does not seem to have been published; on the whole I follow Folkerts 1970 on this difficult subject. The latter contains editions of (despite the title) both 'Geometria I' and 'Geometria II'.

Muhammad Abu al-Wafa al-Buzjani

On the Buyid dynasty and politics see Robinson 2010 (chapters 8 and 9) as well as Mez 1937 and Kraemer 1992; for the geometry of the period Høyrup 1994 and Berggren 2016.

On al-Wafa the articles in Hockey 2014, *Encyclopedia Iranica* (www.iranicaonline.org) and Fleet et al. are valuable, as are the papers in Qurbani and Haydarniya 2002. The 'obvious mistake' line is quoted in the *DSB*. Particular aspects of his work and its transmission are discussed in Kennedy 1984 and Raynaud 2012. For French and German translations of the *Geometric Constructions* see Woepcke 1855, Suter 1922 and Chavoshi 2010, as well as the extracts in English in Katz 2007 (p. 612 for the 'number of geometers' passage) and, more briefly, in Berggren 1986. Raynaud 2012 also contains a complete list of the problems in the treatise, and Sarhangi 2008 an illuminating discussion of parts of its contents. Euclid's *Divisions* is restored in Archibald 1915.

Özdural (1995, 2000) discusses the tradition of meetings between geometers and artisans. A summary of various historians' views on Islamic architectural patterning and its relationship to mathematics appears in Bier 2008 and 2009; the latter has a helpful bibliography of other works on the subject; Kheirandish 2008 is also useful on this subject.

Lady Geometria

Depictions of Euclid are briefly surveyed in the article about him in *BNP*. On Chartres and its sculptural scheme I have used Katzenellenbogen 1959; the photographs in Houvet 1919 are invaluable. I have also used Knitter 2000.

On the Artes and their depiction, including all of the examples I mention, see Verdier 1969, Evans 1978, Zaitsev 1999 and Palmer 2002, as well as several of the images and discussions in Murdoch 1984; the anonymous German manuscript mentioned is described at p. 99.

MS Ashmole 304 is reproduced in full and described in both digital. bodleian.ox.ac.uk and Iafrate 2015; see also Burnett 1977.

Piero della Francesca

The relatively well-trodden ground of this chapter is surveyed for instance in Gamwell 2016 and Kemp 2016; relevant discussions are also in Rose 1975 and Richter 1995.

From the vast amount that has been written about Piero della Francesca I have found Wood 2002 of particular help among general works. On his mathematics Davis 1977 is enormously important, as are the various publications of Field (1996, 1997, 1988) culminating in 2005 which remains the starting point for serious discussion of the subject. Also useful are Elkins 1987 and Folkerts 2006 (chapter X). Francesca's own treatises are all available in modern editions, as is his Archimedes.

On the *Flagellation* in particular the conclusions of Wittkower and Carter 1953 about its perspective have not been substantially improved upon, although Geatti and Fortunati 1992 is also valuable. I have also used Lavin 1972 and Mercier 2017 (the latter has the list of 29 identifications I mention) as well as the discussions of this painting in the more general works noted above, particularly Field 2005.

On Pacioli, Baldasso 2010 and Baldasso and Logan 2017 are levelheaded and informative readings of the much-discussed portrait; Folkerts 2006 (chapter XI) is also valuable.

On the longer history of perspective, Andersen 2007 is unsurpassed as a technical survey; Brownson 1981 is a useful discussion of the Euclidean connection and Raynaud 2010 has helpful technical information; I have also used Belting 2011 and Edgerton 1991. For the later history of the Archimedean solids see Field 1997.

Euclid Speidell

'A new sort of intellectual' is from Oosterhoff 2018, p. 5. Nathaniel Denew's relationship with Euclid Speidell is recorded in Speidell 1686, fols A2r–A3r. Further information about the Speidells is derived from their own works and the International Genealogical Index, as well as from Taylor 1954, where Robert Hooke's comment on him is mentioned. Quotes from John Speidell are variously from Speidell 1616 (pp. 71, 106), 1627 (fol. A3v) and 1628 (fol. A4r). The manuscript autobiography mentioned in the text is in the Lincolnshire Archives, MON 7/21; my thanks for Boris Jardine for this reference and for sight of his unpublished photographs and transcription.

Isaac Newton

The standard source for Newton's life is Westfall 1980; his reading of Euclid is discussed in more detail in Whiteside 1967. His use of his books is addressed in Harrison 1978 and Mandelbrote 2001. Probably the clearest discussion of the chronology of his reading of Descartes and Euclid is that of de Moivre quoted by Whiteside, but the autobiographical evidence is inconsistent and can be interpreted in more than one way.

The quote from Isaac Barrow is from his 1660, fol. 2r. The mathematical methods of Newton's *Principia* are discussed in Guicciardini 1999 and 2009.

On the wider culture of mathematical reading and annotation in this period see Wardhaugh (2015, 2020a, 2020c) as well as the essays in Beeley, Nasifoglu and Wardhaugh 2020. Some of the information in this chapter derives from study of Newton's annotated books themselves in Trinity College, Cambridge; his copy of Barrow's Euclid has shelfmark NQ.16.201[1].

Mary Fairfax

Mathematics and mental illness in this period is discussed in Jenkins 2010 and the sources cited there; see also Jenkins 2007, 2008 and Wardhaugh 2012. On Dürer see Palmer 2002 and Kemp 2016. The wider context of mathematical education in the period is addressed for instance in Richards 1988 and Warwick 2003.

'Not the fool that sticks' is quoted in Heath 1926, vol. 1, p. 415, and Bellman's complaint in *BNP s.v.* 'Euclid'. Stanhope's words are from his 1774, letter 9. De Staël (from *De l'Allemagne* vol. 1, p. 178), Paley and 'Regrets of a Cantab' (from the 1825 London Magazine) are quoted from Jenkins 2010, pp. 92, 95, 96.

For Mary Somerville's life the key source is Somerville 1873, whence I draw my quotations 'seldom read anything' (p. 7), 'not a favourite' (p. 42), 'a monthly magazine' (p. 45), 'the foundation not only of perspective' (p. 49) and the opening passage 'I had to take part' (p. 53). See also Patterson 1969, Neeley 2001 and Lamprecht 2015. 'Whatever difficulty we might experience' is from the obituary in the *Morning Post*, 2 December 1872, quoted in Patterson 1969, p. 311.

François Peyrard

On Peyrard himself see Langins 1989 ('tears of blood', p. 3); the biographical information in Michaud 1843 does not seem certainly reliable. His own account of his work on ancient mathematics in the prefaces to Peyrard 1804 and 1814 helps to complete the picture. The marginal annotation about 'the new edition' is quoted in Vitrac 1990, vol. 1, p. 45; 'the pure text of Euclid' is from Peyrard 1814, vol. 1, p. xxiii; 'calumnies' and 'persecutions' appear in Peyrard 1814, vol. 3, p. xi.

On textual criticism in this period more generally, Reynolds and Wilson 1968/2013 and particularly Pfeiffer 1976 have useful summaries; the history of New Testament criticism is summarised in Metzger 1964 and in Aland and Aland 1987; for the Homeric question the article under that title in *BNP* is a helpful summary.

On the subsequent history of the Euclidean text, the best recent summaries are Rommevaux et al. 2001 and Vitrac 2012; further details are in Vitrac 1990. The accounts in Knorr 1996 and 2001 contain much that is of value.

Nicolai Ivanovich Lobachevskii

On Lobachevskii's life see Vasil'ev 1894, Vucinich 1962 and Manturov 1993 ('a clear thread': p. 9) as well as the useful summary in the *DSB*. Among his own writings, (1891) is probably the most helpful introduction in English: the 'momentous gap' appears at p. 11. The content of non-Euclidean geometry is discussed in Poincaré 1905 and the classic account Coxeter 1957. The older standard history of non-Euclidean geometry Bonola 1912 should be read in conjunction with Gray 1979, 1989 and 2007. 'In triangles whose sides are attainable' is quoted from the latter, p. 122 (from Lobachevskii's *Untersuchungen*); 'it is unlikely' is from p. 129.

Maggie and Tom

Mathematics in Victorian fiction is the subject of Jenkins 2007 and Bayley 2009; on Victorian mathematics more generally see Richards 1988, Jenkins 2008 and Flood, Rice and Wilson 2011. Coleridge and Whewell are quoted from Jenkins 2007, p. 159; 'a dose of mathematics every day' is from Haight 1954, vol. 1, p. 321, 'to lose the power of learning' from vol. 9, p. 293; both quoted in Ball 2015, p. 217.

In addition to the various biographies of Eliot, context for the use of mathematics in her novels is provided by Baker 1977 and Harris and Johnston 1998. Studies dealing with mathematics and/or education in *The Mill on the Floss* and other works include Jones 1995, Jenkins 2008, Ball 2015, 2016 (also Lee 2016) and Dimitriadis 2018.

Simson in Urdu

The edition described is Atmaram 1884. British mathematical textbooks in India are the subject of Aggarwal 2006 and 2007; on the more general subject of education in British India, Sen 1991 is most helpful (p. 199 for 'in the junior scholarship examination'), alongside sources such as Kerr 1852. The activities of British publishers in India are studied by Chatterjee, for instance 2002 and 2006; the quotation is from 2002, p. 153.

On the more general context of British India I have consulted especially Bayly 1988, and on science in British India Arnold 2000; also Dubow 2013 and Chemla 2012, especially chapter 5. Euclid in India before the British is surveyed in De Young 1995.

On the editions of Simson, Playfair and Todhunter – in addition to the editions themselves – see Barrow Green 2006 and Ackerberg-Hastings 2002; Heath 1926 has some helpful information. 'No rival has ever yet risen up' is from Brougham 1845, p. 492; Sylvester is quoted from Moktefi 2011, p. 326.

His modern rivals

On the 'Euclid debate', Dodgson 1879/1885 provides quite a wide-ranging summary of material printed up to that date (the opening quotation is from the start of scene 2); Brock 1975 and Richards 1988 are valuable studies, and Moktefi 2011 (see also 2007) brings the story up to the early twentieth century; Sylvester is once again quoted from Moktefi 2011, p. 326, Perry from p. 334. De Morgan, writing in the *Companion to the Alamanc* for 1849, is quoted in Heath 1926, vol. 1, p. v; 'so bad' is from Russell 1937, p. 405. Quotations from Mueller are from (1981). Wider context is provided by Cajori 1910, Barrow-Green and Gray 2006 and Jenkins 2008.

Thomas Little Heath

I am not aware of secondary literature on Thomas Heath beyond Wardhaugh 2016 and the obituaries and appreciations cited there, notably obituaries in the London *Times*, the *Proceedings* of the British Academy and the Royal Society's *Obituary Notices* (Thompson 1941), and an article in the *Dictionary of National Biography* revised for the *Oxford Dictionary of National Biography*, whence his working habits are gleaned. 'Sound and careful' is from Thompson 1941, p. 413, 'most learned and industrious' p. 409. On the treasury during Heath's period there see Peden 2000; 'the saving of candle-ends' is quoted on p. 6, 'angular and pedantic' on p. 37.

These must be supplemented by what can be gleaned from contemporary reviews and from the prefaces to Heath's own works. From Heath 1920, p. v: 'he is immortal', 'elementary geometry is Euclid', and 'generation after generation'. From Heath 1921: 'mathematics in short is a Greek science' (p. v) and 'if one would understand the Greek genius fully' (p. vi). From Heath 1926, vol. 1: 'I venture to suggest' (p. 418), 'the algebraical method' (p. 373), 'body of doctrine' (p. v) and 'one of the noblest monuments' (p. vii). From Heath 1931, p. 1, 'in the centuries which have since elapsed'. And from Heath 1932b 'the genuine text of Euclid', 'no alternative to the Elements', 'any intelligent person' and 'qualified readers'.

'The true con amore spirit' is from Smith 1909, pp. 387–8; 'in love with geometry' from Aubrey's *Brief Lives*. Wharton is quoted from the frontispiece of Leeke and Serle 1661.

Max Ernst

By far the most detailed study of Ernst's *Euclid* is Lücke-David 1994. The 'Alice' reading originates with Waldberg 1958 (see also Quinn 1977); the 'Rosicrucian' with Hopkins 1992.

The comprehensive catalogue Spies 1975 (vol. 5, p. 106) lists some further literature; of the general works on Ernst, Quinn 1977 has probably the fullest series of reproductions of the works related to *Euclid*, but compare Schneede 1972. I have also consulted Gatt 1970, di san Lazzaro 1971, Diehl 1973, Larkin 1975, Turpin 1979 and Bischoff 2003.

Euclidean designs

On Crockett Johnson see Nel 2012 (from which 'made diagrams', p. 197) as well as www.k-state.edu/english/nelp/purple. On his geometric paintings see Johnson 1972, Stroud 2008, Cawthorne and Green 2009 ('romantic tributes') and Kidwell 2013 ('a kind of imposing thing'), as well as the reproductions and commentary at americanhistory.si.edu/collections/object-groups/mathematical-paintings-of-crockett-johnson. 'A certain cool insouciance' is from Michael Benedikt writing in *Art News*, quoted in Nel 2012, p. 192.

The most complete biography of Oliver Byrne appears to be Hawes and Colpas 2015, whence 'all of my books' and 'I do not introduce colours'. Further information is from www.kroneckerwallis.com, www-history.mcs.st-and.ac.uk, www.c82.net (whence Rougeux's self-description) and math.ubc.ca/~cass/Euclid/byrne.html.

On Elsa Schiaparelli see in particular Schiaparelli 1954, Blum 2003 and Secrest 2014. The 'madder and more original' line is from an article in *Time* 1934. On Madeleine Vionnet I have used Demornex and Canino 1991 and Kamitsis 1996; 'un-European modernity' appeared in the *New Yorker* in 1932, by Janet Flanner.

On Monir Shahroudy Farmanfarmaian the principal printed works are Farmanfarmaian and Houshmand 2007, Obrist and Marta 2011 and Marta 2015. See also Stein 2012. 'I would make a drawing first', 'I asked my master craftsman' and 'it's not a very intellectual process' are from Marta 2015, pp. 54, 96, 102.

Lambda

The two proposed missions which combined to make Euclid are described in Cimatti *et al.* 2009 and Refregier *et al.* 2009. Laureijs *et al.* 2011 describes the outcome of the mission definition study, and mission newsletters for 2012–17 are archived at euclid.cnes.fr/en/newsletters. Updated information about the Euclid space mission appears on the websites of many of the agencies and institutions participating in the Euclid Consortium; see in particular sci.esa.int/euclid/, www.euclid-ec.org and directory.eoportal.org/web/eoportal/satellite-missions/e/euclid.

The 'Euclid' tractors can be found advertised in various issues of the *Illustrated London News* from 1952 to 1960.

Select Bibliography

van der Aa, A.J., *Biographisch woordenboek der Nederlanden* (Haarlem, 1852–78).

Abattouy, Mohammed, 'Greek Mechanics in Arabic Context: Thabit ibn Qurra, al-Isfizari and the Arabic Traditions of Aristotelian and Euclidean Mechanics', *Science in Context* 14 (2001), 179–247.

Acerbi, Fabio, 'Drowning by Multiples. Remarks on the Fifth Book of Euclid's Elements, with Special Emphasis on Prop. 8', *Archive for History of Exact Sciences* 57 (2003) 175–242.

Ackerberg-Hastings, Amy, 'Analysis and Synthesis in John Playfair's *Elements of Geometry*', *British Journal for the History of Science* 35 (2002), 43–72.

Aggarwal, Abhilasha, 'British Higher Education in Mathematics for and in India, 1800–1880' (D.Phil. dissertation, Middlesex University, 2006).

Aggarwal, Abhilasha, 'Mathematical Books for and in India in the Nineteenth Century', *Bulletin of the British Society for the History of Mathematics* 22 (2007), 11–21.

Aghayani-Chavoshi, J., *Ketâb al-Nejârat. (Sur ce qui est indispensable aux artisans dans les constructions géométriques)* (Tehran, 2010).

Aland, Kurt and Barbara Aland, *The Text of the New Testament: An*

Introduction to the Critical Editions and to the Theory and Practice of Modern Textual Criticisms (Grand Rapids, MI, 1987).

Albree, Joe and Scott H. Brown, '"A Valuable Monument of Mathematical Genius": *The Ladies' Diary* (1704–1840)', *Historia Mathematica* 36 (2009), 10–47.

Aletta, A., 'Su Stephano, copista di Areta', *Rivista di Studi Bizantini e Neoellenici* 41 (2004), 73–93.

Alverny, Marie-Thérèse, *La transmission des textes philosophiques et scientifiques au moyen age* (Aldershot, 1994).

Andersen, K., *The Geometry of an Art* (New York, 2007).

Archibald, Raymond Clare, *Euclid's Book on Divisions of Figures = Peri diaireseon biblion: With a restoration based on Woepcke's text and on the* Practica geometriae *of Leonardo Pisano* (Cambridge, 1915).

Arnold, David (ed.), *The New Cambridge History of India III.5: Science, Technology and Medicine in Colonial India* (Cambridge, 2000).

Artmann, Benno, *Euclid: The Creation of Mathematics* (New York, 1999).

Asper, Markus, 'The Two Cultures of Mathematics in Ancient Greece', in Robson and Stedall 2009, 107–32.

Atmaram, *Tahrir-i Uqlidis* (Mathura, 1884).

Bagnall, R.S., *Reading Papyri, Writing Ancient History* (London, 1995).

Baker, William (ed.), *The George Eliot–George Henry Lewes Library: An Annotated Catalogue of Their Books at Dr. Williams's Library, London* (Garland, 1977).

Baldasso, Renzo, 'La stampa dell'editio princeps degli *Elementi di Euclide* (Venice: Erhard Ratdolt, 1482)', in Craig Kallendorf and Lisa Pon (eds), *The Books of Venice/Il Libro Veneziano* (Venice, 2009), 61–100.

Baldasso, Renzo, 'Portrait of Luca Pacioli and Disciple: A New, Mathematical Look', *Art Bulletin* 92 (2010), 83–102.

Baldasso, Renzo, 'Printing For The Doge: On the first quire of the first edition of the *Liber Elementorum Euclidis*', *La Bibliofilía* 115 (2013), 525–52.

Baldasso, Renzo and John Logan, 'Between the Golden Ratio and a Semiperfect Solid: Fra Luca Pacioli and the Portrayal of Mathematical Humanism', in Ingrid Alexander-Skipnes (ed.), *Visual Culture and Mathematics in the Early Modern Period* (New York, 2017), 130–49.

Ball, Derek, 'Thick-rinded Fruit of the Tree of Knowledge': Mathematics education in George Eliot's novels', *Bulletin of the British Society for the History of Mathematics*, 30 (2015), 217–26.

Ball, Derek, 'Mathematics In George Eliot's Novels', (D.Phil. dissertation, University of Leicester, 2016).

Barbera, André, *The Euclidean Division of the Canon: Greek and Latin sources new critical texts and translations on facing pages, with an introduction, annotations, and indices verborum and nominum et rerum* (Lincoln, NE, 1991).

Barker, Andrew, 'Methods and Aims in the Euclidean *Sectio Canonis*', *Journal of Hellenic Studies* 101 (1981), 1–16.

Barker, Andrew, *Greek Musical Writings. Volume 1, The musician and his art* (Cambridge, 1984).

Barker, Andrew, *Greek Musical Writings. Volume 2, Harmonic and acoustic theory* (Cambridge, 1989).

Barker, Andrew, *The Science of Harmonics in Classical Greece* (Cambridge, 2007).

Barrow-Green, June, 'Much Necessary for All Sortes of Men': 450 years of Euclid's *Elements* in English', *Bulletin of the British Society for the History of Mathematics* 21 (2006), 2–25.

Barrow-Green, June and Jeremy Gray, 'Geometry at Cambridge, 1863–1940', *Historia Mathematica* 33 (2006), 315–56.

Barrow, Isaac, *Euclide's Elements; The whole Fifteen Books compendiously Demonstrated By Mr. Isaac Barrow Fellow of Trinity Colledge in Cambridge. And Translated out of the Latin* (London, 1660).

Bayley, Melanie, 'Mathematics and Literature in Victorian England' (D.Phil. dissertation, Oxford, 2009).

Bayly, C.A. (ed.), *The New Cambridge History of India II.1: Indian society and the making of the British Empire* (Cambridge, 1988).

Becker, Oskar, 'Die Lehre vom Geraden und Ungeraden im neunten Buch der Euklidischen Elemente', *Quellen und Studien zur Geschichte der Mathematik, Astronomie, und Physik B* 3 (1936), 533–53.

Beeley, Philip, '"A designe Inchoate". Edward Bernard's planned edition of Euclid and its scholarly afterlife in late seventeenth-century Oxford', in Beeley, Nasifoglu and Wardhaugh 2020.

Beeley, Philip, Yelda Nasifoglu and Benjamin Wardhaugh (eds), *Reading Mathematics in Early-Modern Europe: Studies in the production, collection, and use of mathematical books* (New York, 2020).

Belting, Hans, tr. Deborah Lucas Schneider, *Florence and Baghdad: Renaissance art and Arab science* (Cambridge, MA, 211).

ben Gershom, Levi, ed. Menachem Kellner, *Commentary on Song of Songs* (New Haven, 1998).

Berggren, J.L., *Episodes in the Mathematics of Medieval Islam* (New York, 1986).

Berggren, J.L. and Glen Van Brummelen, 'Al-Kuhi's Revision of Book I of Euclid's *Elements*', *Historia Mathematica* 32 (2005), 426–52.

Berggren, J.L. and Robert S.D. Thomas, *Euclid's* Phaenomena: *A translation and study of a hellenistic treatise in spherical astronomy* (Providence, RI, 2006).

Berkeley, George, *The Analyst; or, a discourse Addressed to an Infidel mathematician.* (London, 1734).

Bier, Carol, 'Art and Mithal: Reading Geometry as Visual Commentary', *Iranian Studies* 41 (2008), 491–509.

Bier, Carol, 'Number, Shape, and the Nature of Space: Thinking through Islamic art', in Robson and Stedall 2009, 827–52.

Billingsley, Henry, *The Elements of Geometrie of the most auncient Philosopher Euclide of Megara* (London, 1570).

Bischoff, Ulrich, *Max Ernst 1891–1976: Beyond painting* (Cologne, 2003).

Blum Rudolf, tr. Hans H. Wellisch, *Kallimachos: The Alexandrian Library and the origins of bibliography* (Madison, WI, 1991).

Blum, Dilys, *Shocking! The art and fashion of Elsa Schiaparelli* (Philadelphia, 2003).

Blumenthal, H., 'Marinus' Life of Proclus: Neoplatonist Biography', *Byzantion* 54 (1984), 469–94.

Bonfante, Larissa and Robert Chipok, *The Plays of Hrotswitha of Gandersheim* (Mundelein, 2013).

Bonola, R., tr. H.S. Carslaw, *History of Non-Euclidean Geometry* (Chicago, 1912).

Bréard, Andrea, '*Imagined Civilizations: China, the West, and Their First Encounter* by Roger Hart [book review]', *Mathematical Intelligencer* 38 (2016), 80–2.

Brentjes, Sonja, 'Varianten einer Haggag-Version von Buch II der *Elemente*' in Folkerts and Hogendijk 1993, 47–67.

Brentjes, S., 'An Exciting New Arabic Version of Euclid's *Elements*. Ms Mumbai, Mulla Fīrūz R.I.6', *Revue d'histoire des mathématiques* 12 (2006), 169–97.

Brentjes, S. 'Euclid's *Elements*, Courtly Patronage and Princely Education', *Iranian Studies* 41 (2008), 441–63.

Brock, W.H., 'Geometry and the Universities: Euclid and his modern rivals, 1860–1901', *History of Education* 4 (1975), 21–35.

Brougham, Henry, *Lives of Men of Letters and Science, who flourished in the time of George III* (London, 1845).

Brownson, C.D., 'Euclid's *Optics* and its Compatibility with Linear Perspective', *Archive for History of Exact Sciences* 24 (1981), 165–94.

Bulmer-Thomas, Ivor, 'Euclid and Medieval Architecture', *Archaeological Journal* 136 (1979), 135–50.

Burgess, Henry E., 'Hroswitha of Gandersheim: A study of the author and her works' (M.A. dissertation, Montana State University, 1957).

Burnett, C., 'What Is the Experimentarius of Bernardus Silvestris?: A Preliminary Survey of the Material', *Archives d'histoire doctrinale et littéraire du moyen âge* 44 (1977) 79–125

Burnett, C. (ed.), *Adelard of Bath: An English scientist and Arabist of the early twelfth century* (London, 1987).

Burnett, C., 'Ocreatus', in Folkerts and Hogendijk 1993, 69–78.

Burton, Harry Edwin, 'The *Optics* of Euclid', *Journal of the Optical Society of America* 35 (1945), 357–72.

Busard, H.L.L., *The Translation of the Elements of Euclid from the Arabic into Latin by Hermann of Carinthia (?), books VII–XII* (Amsterdam, 1977).

Busard, H.L.L (ed.), *The first Latin translation of Euclid's* Elements *commonly ascribed to Adelard of Bath: books I–VIII and books X.36–XV.2* (Toronto, 1983).

Busard, H.L.L (ed.), *The Latin Translation of the Arabic Version of Euclid's* Elements *Commonly Ascribed to Gerard of Cremona* (Leiden, 1984).

Busard, H.L.L and Menso Folkerts (eds), *Robert of Chester's (?) Redaction of Euclid's* Elements, *the So-Called Adelard II Version* (Basel, 1992).

Busard, H.L.L (ed.), *A Thirteenth-Century Adaptation of Robert of Chester's Version of Euclid's* Elements (Munich, 1996).

Busard, H.L.L (ed.), *Johannes de Tinemue's Redaction of Euclid's* Elements, *The So-Called Adelard III Version* (Stuttgart, 2001).

Busard, H.L.L, *Campanus of Novara and Euclid's Elements* (Stuttgart, 2005).

Butler, Alfred J. and P.M. Fraser, *The Arab Conquest of Egypt and the Last Thirty Years of the Roman Dominion* (2nd edition, Oxford, 1978).

Butzmann, Hans, *Corpus agrimensorum Romanorum. Codex arcerianus A der Herzog-August-Bibliothek zu Wolfenbüttel (Cod. Guelf. 36.23A)* (Leiden, 1970).

Byrne, L., 'The Geometrical Method in Spinoza's *Ethics*', *Poetics Today* 28 (2007), 443–74.

Cajori, Florian, 'Attempts Made During the Eighteenth and Nineteenth Centuries to Reform the Teaching of Geometry', *American Mathematical Monthly* 17 (1910), 181–201.

Cameron, Alan and Jacqueline Long, *Barbarians and Politics at the Court of Arcadius* (Berkeley, 1993).

Campbell, Brian, *The Writings of the Roman Land Surveyors: Introduction, text, translation and commentary* (London, 2000).

Carder, James N., *Art Historical Problems of a Roman Land Surveying Manuscript, the Codex Arcerianus A, Wolfenbüttel* (New York, 1978).

Carlebach, Joseph, *Lewi ben Gerson als mathematiker: ein beitrag zur geschichte der mathematik bei den Juden* (Berlin, 1910).

Carlson, Marla, 'Impassive Bodies: Hrotsvit Stages Martyrdom', *Theatre Journal* 50 (1998), 473–87.

Cawthorne, Stephanie and Judy Green, 'Harold and the Purple Heptagon', *Math Horizons* 17 (2009), 5–9.

Černý, Jaroslav, *Paper & Books in Ancient Egypt: An inaugural lecture delivered at University College, London, 29 May, 1947* (London, 1952).

Chartier, Roger, *The Order of Books: Readers, Authors and Libraries in Europe Between the 14th and 18th Centuries* (Stanford, CA, 1985).

Chatterjee, Rimi B., 'Macmillan in India', in E. James (ed.), *Macmillan: A publishing tradition* (London, 2002), 153–69.

Chatterjee, Rimi B., *Empires of the Mind: A history of Oxford University Press in India under the Raj* (Oxford, 2006).

Chemla, K., 'Que signifie l'expression "mathématiques européeennes" vue de Chine?', in Catherine Goldstein, Jeremy Gray and Jim Ritter (eds), *L'Europe mathématique: histoires, mythes, identités* (Paris, 1996), 219–45.

Chemla, K. (ed.), *The History of Mathematical Proof in Ancient Traditions* (Cambridge, 2012).

Cimatti, A., *et al.*, 'SPACE: The spectroscopic all-sky cosmic explorer', *Experimental Astronomy* 23 (2009), 39–66.

Claessens, Guy, 'Proclus: Imagination as a Symptom', *Ancient Philosophy* 32 (2012), 393–406.

Coffman, George R., 'A Note on Saints' Legends', *Studies in Philology* 28 (1931), 580–6.

Corry, Leo, 'Geometry and Arithmetic in the Medieval Traditions of Euclid's *Elements*: a View from Book II', *Archive for History of Exact Sciences* 67 (2013), 637–705.

Costa, Shelley, 'The Ladies' Diary: Society, Gender and Mathematics in England, 1704–1754' (Ph.D. Dissertation, Cornell University, 2000).

Costa, Shelley, 'The "Ladies' Diary": Gender, mathematics, and civil society in early-eighteenth-century England', *Osiris* 17 (2002a), 49–73.

Costa, Shelley, 'Marketing Mathematics in Early Eighteenth-Century England: Henry Beighton, Certainty, and the Public Sphere', *History of Science* 40 (2002b), 211–32.

Coulter, Cornelia C., 'The "Terentian" Comedies of a Tenth-Century Nun', *Classical Journal* 24 (1929), 515–29.

Coxeter, H.S.M., *Non-Euclidean Geometry* (3rd edition, Toronto, 1957).

Cuomo, S., *Ancient Mathematics* (London, 2001).

Curley, E.M., *Behind the Geometrical Method: A Reading of Spinoza's Ethics* (Princeton, 1988).

Cuvigny, Hélène, *L'arpentage par espéces dans l'Égypte ptolémaïque d'après les papyrus grecs* (Brussels, 1985).

d'Elia, Pasquale M., *Galileo in China: Relations through the Roman College between Galileo and the Jesuit scientist-missionaries (1610–1640)* (Cambridge, MA, 1960).

D'Ooge, Martin Luther, Frank Egleston Robbins and Louis Charles Karpinski, *Nicomachus of Gerasa: Introduction to arithmetic* (New York, 1926).

Dahan, Gilbert (ed.), *Gersonide en son temps* (Louvain, Belgium, and Paris, 1991).

Davis, M.D., *Piero della Francesca's Mathematical Treatises: The Trattato d'abaco and Libellus de quinque corporibus regularibus* (Ravenna, 1977).

De Dijn, Herman, 'Conceptions of Philosophical Method in Spinoza: Logica and mos geometricus', *Review of Metaphysics* 40 (1986), 55–87.

de Risi, V., 'The Development of Euclidean Axiomatics', *Archive for History of Exact Sciences* 70 (2016), 591–676.

De Young, Gregg, 'The Arabic Textual Traditions of Euclid's *Elements*', *Historia Mathematica* 11 (1984), 147–60.

De Young, Gregg, 'Ishaq ibn Hunayn, Hunayn ibn Ishaq, and the Third Arabic Translation of Euclid's *Elements*', *Historia Mathematica* 19 (1992), 188–99.

De Young, Gregg, 'Euclidean Geometry in the Mathematical Tradition of Islamic India', *Historia Mathematica* 22 (1995), 138–53.

De Young, Gregg, 'Euclidean Geometry in Two Medieval Islamic Encyclopaedias', *al-Masaq* 14 (2002), 47–60.

De Young, Gregg, 'The Arabic Version of Euclid's *Elements* by al-Hajjaj ibn Yusuf ibn Matar: New Light on a Submerged Tradition', *Zeitschrift für Geschichte der arabisch-islamischen Wissenchaften* 15 (2002/2003), 125–64.

De Young, Gregg, 'The Latin Translation of Euclid's *Elements* Attributed to Gerard of Cremona in Relation to its Arabic Antecedents', *Suhayl: Journal for the History of the Exact and Natural Sciences in Islamic Civilization* 4 (2004), 311–83.

De Young, Gregg, 'Book XVI: A Medieval Arabic Addendum to Euclid's *Elements*', *SCIAMVS: Sources and Commentaries in Exact Sciences* 9 (2008), 133–210.

De Young, Gregg, 'Recovering Truncated Texts: Examples from the Euclidean Transmission', in E. Calvo, M. Comes, R. Puig and M. Ruis (eds), *A Shared Legacy: Islamic Science East and West* (Barcelona, 2008), 247–81.

De Young, Gregg, 'The Tahrir Kitab Usul Uqlidis of Nasir al-Din al-Tusi: Its Sources', *Zeitschrift für Geschichte der arabisch-islamischen Wissenschaften* 18 (2009), 1–71.

Dear, Peter, *Discipline & Experience: The mathematical way in the scientific revolution* (Chicago, 1995).

Della Rocca, Michael (ed.), *The Oxford Handbook of Spinoza* (New York, 2018).

Demornex, Jacqueline and Patricia Canino, *Madeleine Vionnet* (London, 1991).

Di San Lazzaro, G. (ed.), *Homage to Max Ernst* [Special issue of the *XXe Siècle Review*] (New York, 1971).

Diehl, Gaston, tr. Eileen Hennessy, *Max Ernst* (New York, 1973).

Diehl, Robert, *Erhard Ratdolt, ein Meisterdrucker des XV. und XVI. Jahrhunderts* (Vienna, 1933).

Dilke, O.A.W., *The Roman Land Surveyors: An introduction to the agrimensores* (Newton Abbott, 1971).

Dimitriadis, Kimberley, 'Telescopes in the Drawing-Room: Geometry and Astronomy in George Eliot's *The Mill on the Floss*', *Journal of Literature and Science* 11 (2018), 1–19.

Djebbar, Ahmed, 'Quelques Commentaires sur les Versions arabes des *Eléments* d'Euclide et sur leur Transmission à l'Occident Musulman', in Menso Folkerts (ed.), *Mathematische Probleme im Mittelalter. Der lateinische und arabische Sprachbereich* (Wiesbaden, 1996), 91–114.

Djebbar, A., 'Quelques exemples de scholies dans la tradition arabe des Éléments d'Euclide', *Revue d'Histoire des Sciences* 56 (2003), 293–321.

Dodgson, Charles Lutwidge, *Euclid and His Modern Rivals* (London, 1879, 1885).

Dubow, Saul (ed.), *The Rise and Fall of Modern Empires II: Colonial Knowledge* (Farnham, 2013).

Dzielska, Maria, tr. F. Lyra, *Hypatia of Alexandria* (Cambridge, MA, 1995).

Eagles, C.M., 'The Mathematical Work of David Gregory, 1659–1708' (PhD dissertation, University of Edinburgh, 1977).

Edgerton, Samuel Y., *The Heritage of Giotto's Geometry. Art and Science on the Eve of the Scientific Revolution* (Ithaca, 1991).

Edwards, M.J., *Neoplatonic Saints: The lives of Plotinus and Proclus by their students* (Liverpool, 2000).

Eisenstein, Elizabeth, *The Printing Revolution in Early Modern Europe* (2nd edition, Cambridge, 2005).

Eisner, Caroline L., 'Shifting the Focus: Anne Lister as Pillar of Conservatism' *Auto/Biography Studies* 17 (2001), 28–42.

El-Abbadi, Mostafa, *The Life and Fate of the Ancient Library of Alexandria* (Paris, 1990).

Elior, Ofer, 'The Arabic Tradition of Euclid's *Elements* Preserved in the Latin Translation by Adelard of Bath and the Hebrew Translation by Rabbi Jacob', *Historia Mathematica* 45 (2018), 111–30.

Elkins, James, 'Piero della Francesca and the Renaissance Proof of Linear Perspective', *Art Bulletin* 69 (1987), 220–30.

Ellis, Walter M., *Ptolemy of Egypt* (London, 1994).

Encyclopedia Iranica: www.iranicaonline.org.

Endress, Gerhard, 'The Circle of Al-Kindī. Early Arabic Translations from the Greek and the Rise of Islamic Philosophy', in Gerhard Endress, Remke Kruk and H.J. Drossaart Lulofs, (eds), *The Ancient Tradition in Christian and Islamic Hellenism* (Leiden, 1997), 43–76.

Engelfriet, Peter M., *Euclid in China: The genesis of the first Chinese translation of Euclid's Elements, books I–VI (Jihe yuanben, Beijing, 1607) and its reception up to 1723* (Leiden, 1998)

Engelfriet Peter M., 'The Chinese Euclid and its European context', in Catherine Jami and Hubert Delahaye (eds), *L'Europe en Chine: interactions scientifiques, religieuses et culturelles aux XVIIe et XVIIIe siècles* (Paris, 1993), 111–35.

Erskine, Andrew (ed.), *A Companion to the Hellenistic World* (Blackwell, 2003).

Evans, G.R., 'The "Sub-Euclidean" Geometry of the Earlier Middle Ages', *Archive for History of Exact Sciences* 16 (1977), 105–18.

Evans, Michael, 'Allegorical Women and Practical Men: The Iconography of the Artes Reconsidered', *Studies in Church History* 1 (1978), 305–29.

Farmanfarmaian, Monir Shahroudy and Zara Houshmand, *A Mirror Garden* (New York, 2007).

Febvre, Lucien and Henri-Jean Martin, tr. D. Gerard, *The Coming of the Book: The Impact of Printing 1450–1800* (London, 1976; originally published 1958).

Feingold, Mordechai (ed.), *Jesuit Science and the Republic of Letters* (Cambridge, MA, 2002).

Feldman, Seymour, *Gersonides: Judaism within the limits of reason* (Oxford, 2010).

Field, J.V., 'Perspective and the Mathematicians: Alberti to Desargues' in C. Hay (ed.), *Mathematics from Manuscript to Print, 1300–1600* (Oxford, 1988), 236–63.

Field, J.V. 'Piero della Francesca as a Practical Mathematician: The painter as teacher', in Marisa Dalai Emiliani and Valter Curzi (eds), *Piero della Francesca tra arte e scienza* (Venice, 1996), 331–54.

Field, J.V., 'Alberti, the Abacus and Piero della Francesca's Proof of Perspective', *Renaissance Studies* 11 (1997), 61–88

Field, J.V., *Piero Della Francesca: A mathematician's art* (New Haven, CT, 2005).

Fife, Robert Herndon, *Hroswitha of Gandersheim* (New York, 1947).

Fleet, Kate *et al.* (eds), *Encyclopaedia of Islam* (3rd edition, brill.com).

Flood, Raymond, Adrian C. Rice and Robin Wilson, *Mathematics in Victorian Britain* (Oxford, 2011).

Folkerts, M. *"Boethius" Geometrie II* (Wiesbaden, 1970).

Folkerts, Menso, *Essays on Early Medieval Mathematics: The Latin tradition* (Aldershot, 2003).

Folkerts, Menso, *The Development of Mathematics in Medieval Europe: The Arabs, Euclid, Regiomontanus* (Aldershot, 2006).

Fowler, D.H., *The Mathematics of Plato's Academy: A new reconstruction* (Oxford, 1987, 1999).

Frankforter, A. Daniel, 'Hroswitha of Gandersheim and the Destiny of Women', *Historian* 4 (1979), 295–314.

Frantz, Alison, Homer A. Thompson and Ioannes N. Traulos, *Late Antiquity, A.D. 267–700* (Princeton, 1988).

Fraser, P.M., *Ptolemaic Alexandria* (Oxford, 1972).

Freudenthal, G. (ed.), *Studies on Gersonides – A Fourteenth Century Jewish Philosopher-Scientist* (Leiden, 1992).

Freudenthal, Gad, 'Science in the Medieval Jewish Culture of Southern France', *History of Science* 33 (1995), 23–58.

Freudenthal, Gad, *Science in the Medieval Hebrew and Arabic Traditions* (Aldershot, 2005).

Gamwell, Lynn, *Mathematics + Art: A cultural history* (Princeton, 2016).

Garrett, A.V., *Meaning in Spinoza's Method* (Cambridge, 2003).

Garrett, Don (ed.), *The Cambridge Companion to Spinoza* (Cambridge, 1996).

Gatt, Giuseppe, *Max Ernst* (London, 1970).

Geatti, Laura and Luciano Fortunati, 'The Flagellation of Christ by Piero della Francesca: A Study of Its Perspective', *Leonardo* 25 (1992), 361–7.

Gernet, Jacques, *China and the Christian Impact: A conflict of cultures* (Cambridge, 1985).

Geymonat, M., 'Arithmetic and Geometry in Ancient Rome: Surveyors, intellectuals, and poets', *Nuncius* 24 (2009), 11–34.

Giliot, Claude (ed.), *Education and Learning in the Early Islamic World*, (Farnham, 2012).

Glasner, Ruth, *Gersonides: A portrait of a fourteenth-century philosopher-scientist* (Oxford, 2015).

Glucker, John, *Antiochus and the late Academy* (Göttingen, 1978).

Goldstein, Bernard R., 'The Role of Science in the Jewish Community in Fourteenth Century France', *Annals of the New York Academy of Sciences* 314 (1978), 39–49.

Goodrich, L. Carrington and Zhaoying Fang, *Dictionary of Ming biography, 1368–1644* (New York, 1976).

Gossett, Suzanne, 'Blame Not our Author', *Collections of the Malone Society* (1983), 85–132.

Goulding, Robert, *Defending Hypatia: Ramus, Savile, and the Renaissance Rediscovery of Mathematical History* (Dordrecht, 2010).

Grafton, Anthony, *Defenders of the Text: The traditions of scholarship in an age of science, 1450–1800* (Cambridge, MA, 1991).

Grafton, Anthony, *Commerce with the Classics: Ancient Books and Renaissance Readers* (Ann Arbor, MI, 1997).

Grattan-Guinness, I. 'Numbers, Magnitudes, Ratios, and Proportions in Euclid's Elements: How did he handle them?', *Historia Mathematica* 23 (1996), 355–75.

Gray, Jeremy, 'Non-euclidean Geometry — A re-interpretation', *Historia Mathematica* 6 (1979), 236–58.

Gray, Jeremy, *Ideas of Space: Euclidean, non-Euclidean, and relativistic* (Oxford, 1979, 1989).

Gray, Jeremy, *Worlds Out of Nothing: A course in the history of geometry in the 19th century* (London, 2007).

Green, Peter, *Alexander the Great and the Hellenistic Age: A short history* (London, 2008).

Gregory, David, Ευκλειδου τα σωζομενα. *Euclidis quæ supersunt omnia* (Oxford, 1703).

Guicciardini, Niccolò, *Reading the* Principia: *The debate on Newton's mathematical methods for natural philosophy from 1687 to 1736* (Cambridge, 1999).

Guicciardini, Niccolò, *Isaac Newton on Mathematical Certainty and Method* (Cambridge, MA, 2009).

Gutas, Dimitri, *Greek Thought, Arabic Culture. The Graeco-Arabic Translation Movement in Baghdad and Early Abbasid Society (2nd–4th/8th–10th centuries)* (London and New York, 1998).

Haas, Christopher, *Alexandria in Late Antiquity: Topography and social conflict* (Baltimore, 1997).

Haight, Anne Lynn, *Hroswitha of Gandersheim* (New York, 1965).

Haight, Gordon Sherman (ed.), *The George Eliot letters* (London, 1954–78).

Harari, Orna, 'Methexis and Geometrical Reasoning in Proclus' Commentary on Euclid's *Elements*', *Oxford Studies in Ancient Philosophy* 30 (2006), 361–89.

Harris, Margaret and Judith Johnston, *The Journals of George Eliot* (Cambridge, 1998).

Harrison, John, *The Library of Isaac Newton* (Cambridge, 1978).

Hart, Roger, *Imagined Civilizations: China, the West, and their first encounter* (Baltimore, 2013).

Haskins, C.H., *Studies in the History of Mediaeval Science* (Cambridge, MA, 1924).

Haskins, C.H., *The Renaissance of the 12th Century* (Cambridge, MA, 1927).

Hawes, Susan M. and Sid Kolpas, 'Oliver Byrne: The Matisse of Mathematics – Biography 1810–1829', *Convergence* (www.maa.org/

press/periodicals/convergence/oliver-byrne-the-matisse-of-mathematics-biography-1810-1829).

Heath, Thomas Little, *The Thirteen Books of Euclid's* Elements (Cambridge, 1908, 1926).

Heath, Thomas Little, *Euclid in Greek Book I* (Cambridge, 1920).

Heath, Thomas Little, *A History of Greek Mathematics* (Oxford, 1921).

Heath, Thomas Little, *A Manual of Greek Mathematics* (Oxford, 1931).

Heath, Thomas Little, *Greek Astronomy* (London, 1932).

Heath, Thomas Little, 'Introduction' in Isaac Todhunter (ed.), *The Elements of Euclid* (London, 1932).

Heath, Thomas Little, *Mathematics in Aristotle* (Oxford, 1949).

Hirst, Anthony and M.S. Silk, *Alexandria, Real and* Imagined (Aldershot, 2004).

Hobbes, Thomas, *Elementa philosophica de cive* (Amsterdam, 1647).

Hobbes, Thomas, *Examinatio and emendatio mathematicae hodiernae Examinatio & emendatio mathematicæ hodiernæ* (London, 1660).

Hockey, Thomas, *et al.* (eds), *Biographical Encyclopedia of Astronomers* (New York, 2014).

Hogendijk, J.P. 'The Arabic Version of Euclid's *On Divisions'*, Folkerts and Hogendijk 1993, 143–62.

Hölbl, Günther, tr. Tina Saavedra, *A History of the Ptolemaic Empire* (London, 2001).

Hopkins, D., 'Hermetic and Philosophical Themes in Max Ernst's Vox Angelica and Related Works', *Burlington Magazine* 134 (1992), 716–23.

Horowitz, Maryanne Cline (ed.), *New Dictionary of the History of Ideas* (Detroit, 2005).

Houvet, E., *Cathédrale de Chartres: portail occidental ou royal, XIIe siècle* (Chelles, 1919?).

Høyrup, Jens, *In Measure, Number, and Weight. Studies in mathematics and culture* (Albany, 1994).

Hsia, Florence, *Sojourners in a Strange Land: Jesuits and Their Scientific Missions in Late Imperial China* (Chicago, 2010).

Hsia, Florence, 'Roger Hart. *Imagined Civilizations: China, the West, and Their First Encounter* [book review]', *Isis* 106 (2015), 713–16.

Hsia, R.P. *A Jesuit in the Forbidden City, Matteo Ricci, 1552–1610* (Oxford, 2010).

Hughes, Patricia, *The Early Life of Miss Anne Lister and the Curious Tale of Miss Eliza Raine* (Warwick?, 2014).

Hunt, Richard William, *The Survival of Ancient Literature: Catalogue of an exhibition of Greek and Latin classical manuscripts mainly from Oxford libraries* (Oxford, 1975).

Hutton, Charles (ed.), *The Diarian Miscellany* (London, 1775).

Iafrate, Allegra (ed.), *Matthieu Paris, Le moine et le hazard: Bodleian Library, MS Ashmole 304* (Paris, 2015).

Imhausen, Annette, *Mathematics in Ancient Egypt: A contextual history* (Princeton, 2016).

Irigoin, J., 'Survie et renouveau de la littérature antique à Constantinople (IXe siècle)', *Cahiers de civilisation médiévale* 5 (1962), 287–302.

Isham, Thomas, ed. Robert Isham and Walter Rye *The Journal of Thomas Isham, of Lamport: ... from 1st Nov., 1671, to 30th Sept. 1673* (Norwich, 1875).

Isham, Thomas, ed. Gyles Isham, *The Diary of Thomas Isham of Lamport (1658–81)* (Farnborough, 1971).

Jackson, H.J., *Marginalia: readers writing in books* (New Haven, 2001).

Jami, Catherine, 'From Clavius to Pardies: The geometry transmitted to China by Jesuits (1607–1723)', in Federico Masini (ed.), *Western Humanistic Culture Presented to China by Jesuit Missionaries (XVII–XVIII centuries)* (Rome, 1996), 175–99.

Jami, Catherine, Peter M. Engelfriet and Gregory Blue (eds), *Statecraft and Intellectual Renewal in Late Ming China: The Cross-Cultural Synthesis of Xu Guangqi (1562–1633)* (Leiden, 2001).

Jami, Catherine, *The Emperor's New Mathematics: Western Learning and Imperial Authority During the Kangxi Reign (1662–1722)* (Oxford, 2011).

Jardine, Nicholas, 'The Forging of Modern Realism: Clavius and Kepler against the sceptics', *Studies in History and Philosophy of Science* 10 (1979), 141–173.

Jenkins, Alice, *Space and the 'March of Mind': Literature and the Physical Sciences in Britain, 1815–1880* (Oxford, 2007).

Jenkins, Alice, 'George Eliot, Geometry and Gender', in Sharon Ruston (ed.), *Literature and Science: Essays and Studies 2008 for the English Association* (Cambridge, 2008), 72–90.

Jenkins, Alice, 'Mathematics and Mental Health in Early Nineteenth-century England', *Bulletin of the British Society for the History of Mathematics* 25 (2010), 92–103.

Jesseph, Douglas, 'Galileo, Hobbes, and the Book of Nature', *Perspectives on Science* 12 (2004), 191–211.

Johnson, Crockett, 'On the Mathematics of Geometry in My Abstract Paintings', *Leonardo* 5 (1972), 97–101.

Jones, Keith, 'Education, Class and Gender in George Eliot and Thomas Hardy', (D.Phil. dissertation, University of New Hampshire, 1995).

Jones, Matthew L., *The Good Life in the Scientific Revolution: Descartes, Pascal, Leibniz, and the cultivation of virtue* (Chicago, 2006).

Kamitsis, Lydia, *Vionnet* (London, 1996).

Katz, Victor J., *The Mathematics of Egypt, Mesopotamia, China, India, and Islam: A sourcebook* (Princeton, 2007).

Katz, Victor J., *Sourcebook in the Mathematics of Medieval Europe and North Africa* (Princeton, 2016).

Katzenellenbogen, Adolf, *The Sculptural Programs of Chartes Cathedral* (Baltimore, 1959).

Kellner, Menachem, 'Gersonides on the Role of the Active Intellect in Human Cognition', *Hebrew Union College Annual* (1994), 233–59.

Kellner, Menachem, 'Gersonides on the Song of Songs and the Nature of Science', *Journal of Jewish Thought and Philosophy* 4 (1995), 1–21.

Kemp, Martin, *Structural Intuitions: Seeing shapes in art and science* (Charlottesville, 2016).

Kennedy, E.S., 'Applied Mathematics in the Tenth Century: Abu'l-wafa' calculates the distance Baghdad-Mecca', *Historia Mathematica* 11 (1984), 193–206.

Kennedy, Hugh, *The Early Abbasid Caliphate: A political history* (London, 1981).

Kennedy, Hugh, *When Baghdad Ruled the Muslim World: The rise and fall of Islam's greatest dynasty* (Cambridge, MA, 2005).

Kerr, J., *A Review of Public Instruction in the Bengal Presidency from 1835 to 1851* (Calcutta, 1852).

Keyser, Paul T. and Georgia L. Irby (eds), *The Encyclopedia of Ancient Natural Scientists: The Greek tradition and its many heirs* (London, 2008).

Kheirandish, Elaheh, 'Science and Mithal: Demonstrations in Arabic and Persian Scientific Traditions', *Iranian Studies* 41 (2008), 465–89.

Kidwell, Peggy Aldrich, 'The Mathematical Paintings of Crockett Johnson, 1965–1975: An Amateur and His Sources', in Anne Collins Goodyear and Margaret A. Weitekamp (eds), *Analyzing Art and Aesthetics* (Washington, DC, 2013), 198–211.

Knitter, Brian, 'Thierry of Chartres and the West Façade Sculpture of Chartres Cathedral' (MA dissertation, San Jose State University, 2000).

Knobloch, E., 'Sur la vie et l'oeuvre de Christophore Clavius (1538–1612)', *Revue d'Histoire des Sciences* 41 (1988), 331–56.

Knorr, W.R., *The Evolution of the Euclidean Elements: A study of the theory of incommensurable magnitudes and its significance for early Greek geometry* (Dordrecht, 1975).

Knorr, W.R., 'On the Transmission of Geometry from Greek into Arabic', *Historia Mathematica* 10 (1983), 71–8.

Knorr, W.R., *Textual Studies in Ancient and Medieval Geometry* (Boston, 1989).

Knorr, W.R., 'The Wrong Text of Euclid: On Heiberg's text and its alternatives', *Centaurus* 38 (1996), 208–76.

Knorr, W.R., 'On Heiberg's Euclid', *Science in Context* 14 (2001), 133–43.

Koistinen, Olli, *The Cambridge Companion to Spinoza's Ethics* (Cambridge, 2009).

Kraemer, Joel L., *Humanism in the Renaissance of Islam: The Cultural Revival During the Buyid Age* (2nd ed., Leiden, 1992).

Kunitzsch, Paul, '"The Peacock's Tail": On the names of some theorems of Euclid's *Elements*' in Folkerts and Hogendijk 1993, 205–14.

Lamprecht, Elizabeth, 'The Life of Mary Fairfax Somerville, Mathematician and Scientist: a Study in Contrasts', *Michigan Academician* 42 (2015), 1–25.

Langins, Janis, 'Histoire de la vie et des fureurs des François Peyrard, Bibliothécaire de l'École polytechnique de 1795 à 1804 et traducteur renommé d'Euclide et d'Archimède', *Bulletin de la SABIX* 3 (1989), sabix.revues.org/556.

Larkin, David, *Max Ernst* (New York, 1975).

Lassner, Jacob, *The Topography of Baghdad in the Early Middle Ages* (Detroit, 1970).

Lassner, Jacob, *The Shaping of Abbasid Rule* (Princeton, 1980).

Lattis, James, *Between Copernicus and Galileo: Christoph Clavius and the collapse of Ptolemaic cosmology* (Chicago, 1994).

Laureijs, R., *et al.*, *Euclid: Mapping the Geometry of the Dark Universe. Definition study report* (n.p., 2011).

Lavin, Marilyn Aronberg, *Piero della Francesca: The flagellation* (London, 1972).

Lee, Peter M., 'George Eliot and Mathematics', *Bulletin of the British Society for the History of Mathematics* 31 (2016), 154.

Leeke, John and George Serle (eds), *Euclid's Elements of Geometry* (London, 1661).

Lemerle, Paul, tr. Helen Lindsay and Ann Moffatt, *Byzantine Humanism: The first phase: notes and remarks on education and culture in Byzantium from its origins to the 10th century* (Canberra, 1986).

Lévy, Tony, 'Gersonide, commentateur d'Euclide: Traduction annotée de ses gloses sur les Éléments', in Freudenthal 1992, 83–147.

Lévy, Tony, 'Gersonide, le Pseudo-Tusi, et le postulat des parallèles. Les Mathématiques en Hébreu et leurs sources arabes', *Arabic Sciences and Philosophy* 2 (1992), 39–82.

Lévy, Tony, 'The Establishment of the Mathematical Bookshelf of the Medieval Hebrew Scholar: Translations and translators', *Science in Context* 10 (1997), 431–51.

Lévy, Tony, 'Une version hébraïque inédite des *Eléments* d'Euclide', in Danielle Jacquart (ed.), *Les voies de la science grecque. Études sur la transmission des textes de l'Antiquité au dix-neuvième siècle* (Geneva, 1997), 181–239.

Lévy, Tony, 'Les *Eléments* d'Euclide en hébreu (XIIIe–XVIe siècles)', in Ahmed Hasnawi, Abdelali Elamrani-Jamal, and Maroun Aouad (eds), *Perspectives arabes et médiévales sur la tradition scientifique et philosophique grecque* (Leuven and Paris, 1997), 79–94.

Levy, Tony, 'The Hebrew Mathematics Culture (Twelfth–Sixteenth Centuries)' in Gad Freudenthal, (ed.), *Science in Medieval Jewish Cultures* (Cambridge, 2011), 155–71.

Lewis, M.J.T., *Surveying Instruments of Greece and Rome* (Cambridge, 2001).

Lewis, Naphtali, *Papyrus in Classical Antiquity* (Oxford, 1974).

Lewis, Naphtali, *Greeks in Ptolemaic Egypt: Case studies in the social history of the Hellenistic world* (Oxford, 1986).

Leybourn, Thomas (ed.), *The mathematical questions, proposed in the Ladies' diary, and their original answers: together with some new solutions, from its commencement in the year 1704 to 1816* (London, 1817).

Liddington, Jill, 'Anne Lister of Shibden Hall, Halifax (1791–1840): Her Diaries and the Historians', *History Workshop Journal* 35 (1993), 45–77.

Liddington, Jill, *Presenting the Past: Anne Lister of Halifax (1791–1840)* (Hebden Bridge, 1994).

Liddington, Jill, 'Gender, Authority and Mining in an Industrial Landscape: Anne Lister 1791–1840', *History Workshop Journal* 42 (1996), 58–86.

Liddington, Jill, *Female Fortune: land, gender, and authority: the Anne Lister diaries and other writings, 1833–36* (London, 1998).

Lister, Anne, ed. Muriel M. Green, *Miss Lister of Shibden Hall: Selected letters (1800–1840)* (Lewes, 1992).

Lister, Anne, ed. Helena Whitbread, *The Secret Diaries of Miss Anne Lister (1791–1840)* (London, 2010).

Lloyd, G.E.R., *Greek Science after Aristotle* (London, 1973).

Lloyd, Genevieve, *Spinoza and the* Ethics (London, 1996).

Lloyd-Jones, Hugh, Review of *Vanished Library*, *New York Review of Books* (June 14, 1990), 27–9.

Lo Bello, Anthony, *The commentary of Al-Nayrizi on Book I of Euclid's* Elements of geometry, *with an introduction on the transmission of Euclid's* Elements *in the Middle Ages* (Boston, 2003).

Lobachevskii, N.I., tr. George Bruce Halsted, *Geometrical Researches on the Theory of Parallels* (Austin, 1891).

Locke, John, *Posthumous Works* (London, 1706).

Lord, Beth, *Spinoza's* Ethics*: An Edinburgh Philosophical Guide* (Edinburgh, 2010).

Lücke-David, Susanne, *Max Ernst "Euclid": Ein mentales Vexierbild* (Recklinghausen, 1994).

Maclean, Ian, *Scholarship, Commerce, Religion: The Learned Book in the Age of Confessions, 1560–1630* (Cambridge, MA and London, 2012).

MacLeod, Roy M., *The Library of Alexandria: Centre of learning in the ancient world* (London, 2000).

Madan, Falconer, *A summary catalogue of Western manuscripts in the Bodleian Library at Oxford IV: Collections received during the first half of the 19th century* (Oxford, 1897).

Mancha, J.L., 'Levi ben Gerson's astronomical work: Chronology and Christian context', *Science in Context* 10 (1997), 471–93.

Mandelbrote, Scott, *Footprints of the Lion: Isaac Newton at work* (Cambridge, 2001).

Mantas-España, Pedro, 'Was Adelard in Spain? Transmission of knowledge in the first half of the twelfth century', in Charles Burnett and Pedro Mantas-España (eds), *Mapping Knowledge. Cross-Pollination in Late Antiquity and the Middle Ages* (Cordoba, 2014), 195–208.

Manturov, O.V., 'Nikolai Ivanovich Lobachevskii (on the bicentenary of his birth)', *Russian Mathematical Surveys* 48 (1993), 1–13.

Mark, T.C., 'Ordine geometrica demonstrata: Spinoza's Use of the Axiomatic Method', *Review of Metaphysics* 29 (1975), 263–86.

Marlowe, John, *The Golden Age of Alexandria: From its foundation by Alexander the Great in 331 BC to its capture by the Arabs in 642 AD* (London, 1971).

Marrou, Henri Irénée, *Histoire de l'education dans l'antiquité* (Paris, 1948).

Marta, Karen (ed.), *Monir Shahroudy Farmanfarmaian: Works on paper* (Zurich, 2015).

Martzloff, J.-C., 'Le géométrie euclidienne selon Mei Wending', *Historia Scientiarum* 21 (1981), 27–42.

Martzloff, J.-C., 'Clavius traduit en chinois', in Luce Girard (ed.), *Les jésuites à la Renaissance* (Paris, 1995), 309–22.

Masi, Michael (ed.), *Boethian Number Theory: A translation of the De institutione arithmetica (with introduction and notes)* (Amsterdam, 1983).

Matthew, H.C.G. and Brian Harrison (eds), *Oxford Dictionary of National Biography: From the earliest times to the year 2000* (Oxford, 2004).

Mau, Jürgen and Wolfgang Müller, 'Mathematische ostraka aus der Berliner Sammlung', *Archiv für Papyrusforschung* 17 (1960), 1–10.

Mazzio, Carla, 'The Three-dimensional Self: Geometry, Melancholy, Drama', in David Glimp and Michelle R. Warren (eds), *Arts of Calculation: Quantifying thought in early modern Europe* (New York, 2004), 39–65.

Mcdonald-Miranda, Kathryn A., 'Hrotsvit of Gandersheim: Her Works and Their Messages', (M.A. dissertation, Cleveland State University, 2010).

McKeon, Richard, 'Causation and the Geometric Method in the Philosophy of Spinoza', *Philosophical Review* 39 (1930), 178–89, 275–96.

Mercier, Franck, 'Le salut en perspective: Un essai d'interprétation de la

Flagellation du Christ de Piero della Francesca', *Annales: Économies, Sociétés, Civilisations* 72 (2017), 737–71.

Meskens, Ad, *Travelling Mathematics: The fate of Diophantos' arithmetic* (Basel, 2010).

Metzger, Bruce M., *The Text of the New Testament: Its transmission, corruption, and restoration* (Oxford, 1964).

Mez, Adam, tr. Khuda Bakhsh, *The Renaissance of Islam* (London, 1937; repr. New York, 1975).

Michaud, Louis-Gabriel, *Biographie universelle ancienne et moderne* vol. 32 (Paris, 1843).

Moktefi, A., 'How to Supersede Euclid: Geometrical teaching and the mathematical community in nineteenth-century Britain', in Amanda Mordavsky Caleb (ed.), *(Re)Creating Science in Nineteenth-Century Britain* (Newcastle, 2007), 216–29.

Moktefi, Amirouche, 'Geometry: the Euclid debate', in Flood, Rice and Wilson 2011, 320–36.

Montanari, Franco, Stephanos Matthaios and Antonios Rengakos (eds), *Brill's Companion to Ancient Scholarship* (Leiden, 2014).

Moody, E.A. and M. Clagett, *The Medieval Science of Weights* (Madison, 1952).

Mueller Ian, 'Iamblichus and Proclus' Euclid Commentary', *Hermes* 115 (1987), 334–48.

Mueller, Ian, *Philosophy of Mathematics and Deductive Structure in Euclid's* Elements (Cambridge, MA, 1981).

Murdoch, J.E., 'Euclides Graeco-Latinus: A Hitherto unknown medieval Latin translation of the *Elements* made directly from the Greek', *Harvard studies in Classical philology* 71 (1966), 249–302.

Murdoch, J.E., *Antiquity and the Middle Ages [Album of Science]* (New York, 1984).

Nadler, Steven, *Spinoza: A Life* (Cambridge, 1999).

Nadler, Steven, *Spinoza's* Ethics: *An Introduction* (Cambridge, 2006).

Nasifoglu, Yelda, 'Embodied Geometry in Early Modern Theatre', in Justin E.H. Smith (ed.), *Embodiment* (Oxford, 2017), 311–16.

Nasifoglu, Yelda, 'Reading by Drawing. The Changing Nature of Mathematical Diagrams in Seventeenth-Century England', in Beeley, Nasifoglu and Wardhaugh 2020.

Neeley, Kathryn A., *Mary Somerville: Science, illumination, and the female mind* (Cambridge, 2001).

Nel, Philip, *Crockett Johnson and Ruth Krauss: How an Unlikely Couple Found Love, Dodged the FBI, and Transformed Children's Literature* (Jackson, MI, 2012).

Netz, Reviel, 'Classical Mathematics in the Classical Mediterranean', *Mediterranean Historical Review* 12 (1997), 1–24.

Netz, Reviel, 'Deuteronomic Texts: Late antiquity and the history of mathematics', *Revue d'histoire des mathématiques* 4 (1998), 261–88.

Netz, Reviel, *The Shaping of Deduction in Greek Mathematics: A study in cognitive history* (Cambridge, 1999).

Netz, Reviel, 'Counter Culture: Towards a history of Greek numeracy', *History of Science* 40 (2002), 321–52.

Netz, Reviel, 'Greek Mathematicians: A Group Picture', in C.J. Tuplin, T.E. Rihll and Lewis Wolpert (eds), *Science and Mathematics in Ancient Greek Culture* (Oxford, 2002), 196–216.

Netz, Reviel, *From Problems to Equations: A study in the transformation of early Mediterranean mathematics* (Cambridge, 2004).

Netz, Reviel, *Ludic Proof: Greek mathematics and the Alexandrian aesthetic* (Cambridge, 2008).

Netz, Reviel, 'The Texture of Archimedes' Writings: Through Heiberg's veil', in Chemla 2012, 163–205

Neuenschwander, E., 'Die ersten vier Bücher der Elemente Euklids: Untersuchungen über den mathematischen Aufbau, die Zitierweise und die Entstehungsgeschichte', *Archive for History of Exact Sciences* 9 (1973), 325–80.

Neuenschwander, E., 'Die stereometrischen Bücher der Elemente Euklids. Untersuchungen über den mathematischen Aufbau und die Entstehungsgeschichte', *Archive for History of Exact Sciences* 14 (1974), 91–125.

Nicolson, Marjorie Hope and Sarah Hutton (eds), *The Conway Letters: The correspondence of Anne, Viscountess Conway, Henry More, and their friends, 1642–1684* (Oxford, 1992).

Nikulin, Dmitri, 'Imagination and Mathematics in Proclus', *Ancient Philosophy* 28 (2008), 153–72.

Nordqvist, Nils, *Erhard Ratdolt: Euklidestryckaren i Venedig, liturgtryckaren i Augsburg* (Stockholm, 1961).

O'Meara, Dominic J., *Pythagoras Revived: Mathematics and philosophy in late antiquity* (Oxford, 1989).

Obrist, Hans Ulrich and Karen Marta, *Monir Shahroudy Farmanfarmaian: cosmic geometry* (Bologna, 2011).

Olmos, Paula (ed.), *Greek Science in the Long Run: Essays on the Greek Scientific Tradition (4th c. BCE–17th c. CE)* (Newcastle, 2012).

Oosterhoff, Richard J., *Making Mathematical Culture: University and Print in the Circle of Lefèvre d'Étaples* (Oxford, 2018).

Özdural, Alpay, 'Omar Khayyam, mathematicians, and convesazioni with artisans', *Journal of the Society of Architectural Historians* 54 (1995), 54–71.

Özdural, Alpay, 'Mathematics and Arts: Connections between Theory and Practice in the Medieval Islamic World', *Historia Mathematica* 27 (2000), 171–201.

Palmer, William, 'Images of Knowledge: The seven liberal arts and their representation in Medieval and Renaissance art' (MA dissertation, California State University, 2002).

Patterson, Elizabeth C., 'Mary Somerville', *British Journal for the History of Science* 4 (1969), 311–39.

Peden, G.C., *The Treasury and British Public Policy, 1906–1959* (Oxford, 2000).

Pépin, J. and H.D. Saffrey (eds), *Proclus, lecteur et interprète des Anciens* (Paris, 1987).

Perl, Teri, 'The Ladies' Diary or Woman's Almanack, 1704–1841', *Historia Mathematica* 6 (1979), 36–53.

Peyrard, F., *Les Oeuvres d'Euclide* (Paris, 1804).

Peyrard, F., *Les Oeuvres d'Euclide* (Paris, 1814–18).

Pfeiffer, Rudolf, *History of Classical Scholarship from 1300 to 1850* (Oxford, 1976).

Pfeiffer, Rudolf, *History of Classical Scholarship: From the beginnings to the end of the Hellenistic age* (Oxford, 1968).

Poincaré, Henri, tr. W.J. Greenstreet, 'Non-Euclidean Geometries' in *Science and hypothesis* (London, 1905), 42–59.

Porten, Bezalel, *The Elephantine Papyri in English: Three millennia of cross-cultural continuity and change* (Leiden, 1996).

Price, Audrey, 'Pure and Applied: Christopher Clavius's Unifying Approach to Jesuit Mathematics Pedagogy' (D.Phil. dissertation, University of California, 2017).

Proclus, trans. and ed. Glenn R. Morrow, *A Commentary on the First Book of Euclid's* Elements (Princeton, 1970, 1992).

Quinn, Edward, *Max Ernst* (London, 1977).

Qurbani, Abu al-Qasim and Muhsin Haydarniya, *Nabighah-'i Buzjan: guzidah-i maqalat-i Siminar-i Bayn al-Milali-i Abu al-Vafa-yi Buzjani / International Seminar on Abu al-Wafa Buzdjani* (Teheran, 2002).

Raven, James, 'Printing and Printedness', in *The Oxford Handbook of Early Modern European History, 1350–1750* vol. I: *Peoples and Place* (Oxford, 2015).

Raynaud, Dominique, 'Les débats sur les fondements de la perspective linéaire de Piero della Francesca à Egnatio Danti: un cas de mathématisation à rebours', *Early Science and Medicine* 15 (2010), 474–504.

Raynaud, Dominique, 'Abu al-Wafa' Latinus? A study of method', *Historia Mathematica* 39 (2012), 34–83.

Redgrave, Gilbert R., *Erhard Ratdolt and His Work at Venice: A Paper Read before the Bibliographical Society November 20, 1893* (London, 1894).

Refregier, A., *et al.*, 'The DUNE Collaboration', *Experimental Astronomy* 23 (2009), 17–37.

Reynolds, L.D. and N.G. Wilson, *Scribes and Scholars: A guide to the transmission of Greek and Latin literature* (London, 1968; 4th edition Oxford 2013).

Richards, Joan L., *Mathematical Visions: The Pursuit of Geometry in Victorian England* (Boston, 1988).

Richter, Fleur, *Die Ästhetik geometrischer Körper in der Renaissance* (Stuttgart, 1995).

Risk, R. Terry, *Erhard Ratdolt: Master Printer* (Francestown, 1982).

Roberts, Colin H. and T.C. Skeat, *The Birth of the Codex* (London, 1983).

Robinson, Chase F. (ed.), *New Cambridge History of Islam I: The formation of the Islamic world, sixth to eleventh centuries* (Cambridge, 2010).

Robson, Eleanor, 'Influence, Ignorance, or Indifference? Rethinking the relationship between Babylonian and Greek mathematics', *Bulletin of the British Society for the History of Mathematics* 4 (2005), 1–17.

Robson, Eleanor and Jacqueline A. Stedall, *The Oxford Handbook of the History of Mathematics* (Oxford, 2009).

Rommevaux, Sabine, *Clavius: une clé pour Euclide au XVIe siècle* (Paris, 2005).

Rommevaux, Sabine, Ahmed Djebbar and Bernard Vitrac, 'Remarques sur l'Histoire du Texte des Éléments d'Euclide', *Archive for History of Exact Sciences* 55 (2001), 221–95.

Rosán, Laurence Jay, *The Philosophy of Proclus: The final phase of ancient thought* (New York, 1949).

Rose, Paul L., *The Italian Renaissance of Mathematics: Studies on Humanists and Mathematicians from Petrarch to Galileo* (Geneva, 1975).

Rosenthal, Franz, *The Classical Heritage in Islam* (Berkeley, 1975).

Rowlandson, Jane and Andrew Harker, 'Roman Alexandria from the Perspective of the Papyri', in Hirst and Silk 2004, 79–112.

Rudorff, Adolf August Friedrich, Karl Lachmann and Friedrich Blume (eds), *Die Schriften der römischen Feldmesser* (Berlin, 1848).

Russo, L. 'The Definitions of Fundamental Geometric Entities Contained in Book I of Euclid's *Elements*', *Archive for History of Exact Sciences* 52 (1998), 195–219.

Russell, Bertrand, *The Principles of Mathematics* (2nd edition, London, 1937).

Sachs, E. *Die fünf platonischen Körper* (Berlin, 1917).

Saffrey, Henri-Dominique, 'Ἀγεωμέτρητος μηδεὶς εἰσίτω. Une inscription légendaire', *Revue des Études Grecques* 81 (1968), 67–87.

Sarhangi, Reza, 'Illustrating Abu al-Wafa' Buzjani: Flat Images, Spherical Constructions', *Iranian Studies* 41 (2008), 511–23.

Savile, Henry, *Prælectiones tresdecim in principium Elementorum Euclidis* (Oxford, 1621).

Schiaparelli, Elsa, *Shocking Life* (London, 1954).

Schneede, Uwe M., tr. R.W. Last, *The Essential Max Ernst* (London, 1972).

Secrest, Meryle, *Elsa Schiaparelli: A biography* (London, 2014).

Sen, S.N., *Scientific and Technical Education in India 1781–1900* (New Delhi, 1991).

Simonson, S., 'The Missing Problems of Gersonides – A critical edition', *Historia Mathematica* 27 (2000), 243–302, 384–431.

Siorvanes, Lucas, *Proclus: Neo-platonic philosophy and science* (Edinburgh, 1996).

Smith, D.E., 'Heath's Euclid' [review] *Bulletin of the American Mathematical Society* 15 (1909), 386–91.

Smith, Thomas, *Vita clarissimi & doctissimi viri, Edwardi Bernardi, S. Theologiæ Doctoris, Et Astronomiae apud Oxonienses Professoris Saviliani* (London, 1704).

Smolarski, Dennis C., 'The Jesuit Ratio studiorum, Christopher Clavius, and the Study of Mathematical Sciences in Universities', *Science in Context* 15 (2002), 447–57.

Solère, Jean-Luc, 'L'ordre axiomatique comme modèle d'écriture philosophique dans l'Antiquité et au Moyen Âge', *Revue d'Histoire des Sciences* 56 (2003), 323–45.

Somerville, Mary, *Personal Recollections: From early life to old age, of Mary Somerville* (London, 1873).

Speidell, John, *A Geometricall Extraction* (London, 1616).

Speidell, John, *An Arithemeticall Extraction* (London, 1628).

Speidell, John, ed. Euclid Speidell, *An Arithmetical Extraction* (London, 2nd edition, 1686).

Spence, Jonathan D., *The Memory Palace of Matteo Ricci* (New York, 1984; London, 2008).

Spies, Werner, *et al.*, *Max Ernst Oeuvre-Katalog* (Houston, 1975).

St. John, Christopher, *The Plays of Roswitha* (London, 1923).

Stamatēs, Euangelos S., *Euclidis Elementa post I. L. Heiberg edidit E. S. Stamatis* (Leipzig, 1969–1977).

Stanhope, Philip, *Letters written by the late Right Honourable Philip Dormer Stanhope, Earl of Chesterfield, to his son Philip Stanhope, Esq.* (Dublin, 1774).

Steck, Max, *Bibliographia Euclideana: Die Geisteslinies der Tradition in den Editionen der "Element" (Στοιχεια) des Euklid (um 365–300)* (Hildesheim, 1981).

Steenbakkers, Piet, 'The Textual History of Spinoza's *Ethics*' in Koistinen 2009, 26–41.

Steidele, Angela, tr. Derbyshire, Katy, *Gentleman Jack: The biography of Anne Lister, regency landowner, seducer and secret diarist* (London, 2018).

Stein, Donna, 'Monir Shahroudy Farmanfarmaian: empowered by American art: an artist's journey', *Woman's Art Journal* 33 (2012), 3–9.

Stevens, Wesley M., 'Fields and Streams: Language and practice of Arithmetic and Geometry in early medieval schools', in John J. Contreni and Santa Casciani (eds), *Word, Image, Number: communication in the Middle Ages* (Florence, 2002), 113–204.

Stevens, Wesley M., 'Euclidean Geometry in the Early Middle Ages: A preliminary assessment', in Marie-Thérèse Zenner (ed.), *Villard's Legacy: Studies in medieval technology, science and art in memory of Jean Gimpel* (Aldershot 2004), 229–62.

Stroud, James B., 'Crockett Johnson's Geometric Paintings', *Journal of Mathematics and the Arts* 2 (2008), 77–99.

Suter, H., 'Das Buch der geometrischen Konstruktionen des Abū'l Wefa', in *Beiträge zur Geschichte der Mathematik bei den Griechen und Arabern* (1922), 94–109.

Szabó, Árpád, tr. A.M. Ungar, *The Beginnings of Greek Mathematics* (Dordrecht, 1978).

Taylor, E.G.R., *The Mathematical Practitioners of Tudor and Stuart England* (Cambridge, 1954).

Taylor, Thomas, *Sallust, On the gods and the world: and the Pythagoric sentences of Demophilus, translated from the Greek; and five hymns by Proclus, in the original Greek, with a poetical version. To which are added five hymns by the translator* (London, 1793).

Thomas-Stanford, Charles, *Early Editions of Euclid's* Elements (London, 1926).

Thompson, Dorothy J., *Kerkeosiris: An Egyptian village in the Ptolemaic period* (Cambridge, 1971).

Thompson, D.W., 'Sir Thomas Little Heath', *Obituary Notices of Fellows of the Royal Society* 3 (1941), 409–26.

Tobin, Richard, 'Ancient Perspective and Euclid's *Optics*', *Journal of the Warburg and Courtauld Institutes* 53 (1990), 14–41.

Touati, Charles, *La pensée philosophique et théologique de Gersonide* (Paris, 1973).

Tupin, C.J. and T.E. Rigll (eds), *Science and Mathematics in Ancient Greek Culture* (Oxford, 2002).

Turner, E.G., *Greek Papyri: An Introduction* (Oxford 1968).

Turner, E.G., rev. P.J. Parsons, *Greek Manuscripts of the Ancient World* (London, 1987)

Turpin, Ian, *Max Ernst* (Oxford, 1979).

Unguru, Sabetai, 'On the Need to Rewrite the History of Greek Mathematics', *Archive for History of Exact Sciences* 15 (1975), 67–114.

Vasil'ev, Aleksandr Vasil'evich, tr. George Bruce Halsted, *Nicolái Ivánovich Lobachévsky* (Austin, 1894).

Ver Eecke, Paul, *Les commentaires sur le premier livre des Éléments d'Euclide* (Bruges, 1948).

Verdier, P., 'L'iconographie des arts libéraux dans l'art du moyen âge jusqu'à la fin du quinzième siècle', in *Arts libéraux et philosophie au moyen age. Actes du quatrième congres international de philosophie médiévale* (Montréal, 1969), 305–55.

Viljanen, Valtteri, *Spinoza's Geometry of Power* (Cambridge, 2011).

Vioque, Guillermo Galán, 'The Lost Library of Jacques Philippe d'Orville', *Quaerendo* 47 (2017), 132–70.

Vitrac, B., *Les Eléments* (Paris: Presses universitaires de France, 1990–2001).

Vitrac, B., 'La transmission des textes mathématiques: l'exemple des Éléments d'Euclide', in L. Giard and C. Jacob (eds), *Des Alexandries, I., Du livre au texte* (Paris, 2001), 339–55.

Vitrac, Bernard, 'Les scholies grecques aux Éléments d'Euclide / Greek scholia on Euclid's Elements', *Revue d'histoire des sciences* 56 (2003), 275–92.

Vitrac, B., 'À propos des démonstrations alternatives et autres substitutions de preuves dans les Éléments d'Euclide', *Archive for history of exact sciences* 59 (2004), 1–44.

Vitrac, Bernard, 'Euclid' [supplement to the article in *DSB*] (2006).

Vitrac, Bernard, 'Figures du mathématicien et représentations des mathématiques en Grèce ancienne (VIe–IVe s.avant notre ère)' (preprint dated 2008, hal.archives-ouvertes.fr/hal-00454058).

Vitrac, Bernard, 'The Euclidean ideal of proof in The Elements and philological uncertainties of Heiberg's edition of the text', in Chemla 2012, 69–134.

Vitrac, Bernard and A. Djebbar 'Le livre XIV des Éléments d'Euclide: Versions grecques et arabes (première partie)', *Sciamus* 12 (2011), 29–158.

Vitrac, Bernard and A. Djebbar 'Le livre XIV des Éléments d'Euclide: Versions grecques et arabes (seconde partie)', *Sciamus* 13 (2012), 3–156.

Vucinich, Alexander, 'Nikolai Ivanovich Lobachevskii: The Man behind the First Non-Euclidean Geometry', *Isis* 53 (1962, 465–81.

Waldberg, Patrick, *Max Ernst* (Paris, 1958).

Wallace, William, *Galileo and His Sources: The Heritage of the Collegio Romano in Galielo's Science* (Princeton, 1984).

Wallis, Richard T., *Neoplatonism* (2nd edition, London, 1995).

Wardhaugh, Benjamin, *Poor Robin's Prophecies: A curious almanac, and the everyday mathematics of Georgian Britain* (Oxford, 2012).

Wardhaugh, Benjamin, 'Consuming Mathematics: John Ward's *Young Mathematician's Guide* (1707) and its owners', *Journal for Eighteenth-Century Studies* 38 (2015), 65–82.

Wardhaugh, Benjamin, 'Greek Mathematics in English: The work of Sir Thomas L. Heath', in Volker R. Remmert, Martina Schneider and Henrik Kragh Sørenson (eds), *Historiography of mathematics in the 19th and 20th centuries* (Cham, 2016), 109–22.

Wardhaugh, Benjamin, 'Defacing Euclid: Reading and annotating the *Elements of Geometry* in early modern Britain', in Ann, Blair, Paul Duguid, Anja-Silvia Goeing and Anthony Grafton (eds), *Information: A Historical Companion* (Princeton, 2020).

Wardhaugh, Benjamin, 'Euclid in Print, 1482–1704: A catalogue of the editions of the *Elements* and other Euclidean works' (2020: www. readingeuclid.org).

Wardhaugh, Benjamin, 'Rehearsing in the Margins: Mathematical print and mathematical learning', in Alice Jenkins and Robert Tubbs (eds), *Palgrave Companion to Mathematics and Literature* (Palgrave, 2020).

Warwick, Andrew, *Masters of Theory: Cambridge and the rise of mathematical physics* (Chicago, 2003).

Webster, Colin, 'Euclid's *Optics* and Geometrical Astronomy', *Apeiron* 47 (2014), 526–51.

Wedeck, H.E., 'The Humor of a Mediaeval Nun, Hroswitha', *Classical Weekly* 21 (1928), 130–1.

Weitzmann, Kurt, *Illustrations in Roll and Codex: A study of the origin and method of text illustration* (Princeton, 1947).

West, M.L., *Ancient Greek Music* (Oxford, 1992).

Westerink, Leendert Gerrit, *Arethae archiepiscopi Caesariensis Scripta minora* (Leipzig, 1968–1972).

Westfall, Richard S., *Never at Rest: A biography of Isaac Newton* (Cambridge, 1980).

Whitbread, Helena, *I Know My Own Heart: The diaries of Anne Lister (1791–1840)* (London, 1988).

Whiteside, D.T., *The Mathematical Papers of Isaac Newton* (Cambridge, 1967–81).

Wiegand, M. Gonsalva (tr.), *Hroswitha of Gandersheim: The Non-Dramatic Works* (St Louis, 1936).

Wilson, Katharina M., 'The Saxon Canoness: Hrotsvit of Gandersheim', in *Medieval Women Writers* (Manchester, 1984), 30–63.

Wilson, Katharina M., *Hrotsvit of Gandersheim: A florilegium of her works* (Cambridge, 1998).

Wilson, N.G., *Scholars of Byzantium* (London, 1983).

Wittkower, R. and B. Carter, 'The Perspective of Piero della Francesca's Flagellation', *Journal of the Warburg and Courtauld Institutes* 16 (1953), 292–302.

Woepcke, Franz, 'Analyse et extraits d'un recueil de constructions géométriques par Aboûl Wefâ', *Journal asiatique*, 5th series, 5 (1855), 218–56, 309–59.

Wood, Jeryldene, *The Cambridge Companion to Piero della Francesca* (Cambridge, 2002).

Zaitsev, E.A., 'The Meaning of Early Medieval Geometry – From Euclid and surveyors' manuals to Christian philosophy', *Isis* 90 (1999), 522–553.

Zenner, Marie-Thérèse, 'Imaging a Building: Latin Euclid and Practical Geometry', in Contreni and Casciani 2002, 219–46.

Zhang, Qiong, *Making the New World Their Own: Chinese Encounters with Jesuit Science in the Age of Discovery* (Leiden, 2015).

Zhmud, Leonid, 'Plato as "Architect of Science"', *Phronesis* 43 (1998), 211–44.

Zhmud, Leonid, 'Eudemus' History of Mathematics', in M. István Bodnár and William W. Fortenbaugh (eds), *Eudemus of Rhodes* (New Brunswick, 2002), 263–306.

Index

(762), 49; Banu Musa, 53, 55; as
cultural centre, 50–1, 52–5;
culture of meeting and
discussions, 196, 200–1; Indian
and Persian learning, 51, 54;
library ('House of Wisdom'),
52–3, 59; revival under Buyid
dynasty, 193–201; Round City of
al-Mansur, 48, 49–55, 50;
translations of the *Elements* in, 2,
51–2, 54–5; war of succession in
(196 AH, AD 811), 52
Banks, J. Cleaver, 46
Barmakid family, 49
Barrow, Isaac, 231–2, 233, 234
Bartels, Johann Christian Martin,
262
Basilides of Tyre, 29
Beijing, 2, 133, 134, 135, 136–43
Bellman, Carl Mikael, 243
Benedict XII, Pope, 119
Benedictine order, 110–11
Berkeley, Bishop, 152
Bernard (brother of Thierry), 205
Bernard, Edward, 83–7, 89
Billingsley, Henry, 76–7
Blake, William, 242–3, 244
'Blame not our author' (play), 145–50
Bodleian Library (Oxford), 47, 81,
82–3, 86–7, 209
Boethius, 58, 62, 108, 112, 126, 191,
205–6, 207
Bonnycastle, John, 246
'Bourbaki' (group of French
mathematicians), 304
Britain: cultural cachet of Euclid in,
242–3; 'Euclid debate' in
Victorian/Edwardian period,
286–90, 294–9; Euclid's official
demotion in, 289–90, 296–7,
298–9, 316–17; Victorian/
Edwardian education, 269–74,
275–7, 279–80, 286–90, 296–9
see also England; Scotland

British Association for the
Advancement of Science, 290
British East India Company, 280
British Library (London), 207
Brunelleschi, Filippo, 213, 214
Buddhism, 137, 140
Bulgarian khanate, 42
Busby, Richard, 228
Buyid dynasty, 193–201
al-Buzjani, Muhammad Abu al-Wafa,'
193, 194–201, 314
Byrne, Oliver, 309–11
Byzantium: cultural contact with
caliphate, 50–1, 53; founding of
Constantinople (324), 39–40;
Latins take Constantinople
(1204), 57; and New Testament,
256; ninth-century cultural
revival, 42–3; scholarship, 40–6,
256, 296; wars and internal
conflicts, 42, 49

Caesarea, 43
calculus, 234, 247, 263, 275
caliphate: Abbasid dynasty, 2, 48–53;
break-up of empire from later
ninth-century, 194; cultural
contact with Byzantium, 50–1,
53; Fatimid library (Cairo), 57;
rule in Alexandria, 40; sale or
plunder of Islamic libraries, 57;
scholarship and learning, 51–5;
vast world ruled by, 48–9; war of
succession in Baghdad (196 AH,
AD 811), 52; war with
Byzantium, 42
Cambridge University, 230, 231–2,
234, 244–5, 290, 291–2
camera obscura, 118
Campanus of Novara, 63–4, 69, 76
Carolingian empire, 42, 189
Carroll, Lewis: *Euclid and His
Modern Rivals*, 285, 287;
Through the Looking Glass, 302